# Lecture Notes on Coastal and Estuarine Studies

Managing Editors:
Richard T. Barber  Christopher N. K. Mooers
Malcolm T. Bowman  Bernt Zeitzschel

1

# Mathematical Modelling of Estuarine Physics

Proceedings of an International Symposium
Held at the German Hydrographic Institute
Hamburg, August 24–26, 1978

Edited by
J. Sündermann and K.-P. Holz

Springer-Verlag
Berlin  Heidelberg  New York 1980

**Volume Editors**

Jürgen Sündermann
Institut für Meereskunde der Universität Hamburg
Heimhuder Straße 71
D-2000 Hamburg 13
Fed. Rep. of Germany

Klaus-Peter Holz
Lehrstuhl für Strömungsmechanik der Universität Hannover
Callinstraße 32
D-3000 Hannover 1
Fed. Rep. of Germany

ISBN 3-540-09750-3 Springer-Verlag Berlin Heidelberg New York
ISBN 0-387-09750-3 Springer-Verlag New York Heidelberg Berlin

Library of Congress Cataloging in Publication Data. Main entry under title: Mathematical modelling of estuarine physics. (Lecture notes on coastal and estuarine studies ; 1) Bibliography: p. Includes index. 1. Estuaries--Mathematical models--Congresses. I. Sündermann, Jürgen. II. Holz, Klaus-Peter. III. Series. GC96.5.M37 551.46'09 80-11461 ISBN 0-387-09750-3

Printing and binding: Beltz Offsetdruck, Hemsbach/Bergstr.
2131/3140-543210

In honor of
Prof. Walter Hansen
on occasion of his 70th birthday

PREFACE BY THE EDITORS

"Mathematical Modelling of Estuarine Physics" is the first volume
in the new series Lecture Notes on Coastal and Estuarine Studies.
(The prototype for the series was "Oceanic Fronts in Coastal Processes.") This volume was derived from contributions made at the
Symposium on Mathematical Modelling of Estuarine Physics held at
Hamburg, August, 1978. This symposium brought together
numerous oceanographers, coastal engineers, fluid mechanicians,
numerical modellers and other scientists and engineers from several
countries to provide "state-of-the-art" discussions of active
research in modelling the physics of estuarine and coastal regimes.
Hence, the symposium realized several of the distinguishing
criteria of these Lecture Notes: a coastal and estuarine regime
theme; interdisciplinary and international scope; and topicality.
This volume meets several of the criteria of the new series:
peer-review and rapid publication; relatively low price; and
an integrated collection of contributions.

The editors have a few more manuscripts under review, and they
are seeking additional ones. Interested authors are encouraged
to contact one of the editors, who will provide information on
the preparation of a prospectus. The scope of the Lecture Notes
includes biological, chemical, and geological aspects as well
as physical ones, and comprises engineering as well as science
topics. It is essential that the volumes be more integrated than
standard conference proceedings. Also, single-author volumes
are perfectly acceptable. Of some considerable interest is the
rapid dissemination of knowledge in this topic area of global,
societal concern.

THE EDITORS

## PREFACE

Numerical simulations are in world wide use for the investigation
of hydro- and thermodynamic processes in natural waters. In spite
of the fact that great achievements have been brought about in
this field, numerous questions still remain unanswered, with
respect to fundamental formulations as well as to methods of solution. One of the principal objectives of research in our days is
the verification of numerical models.

The Symposium on "Mathematical Modelling in Estuarine Physics" was
held at Hamburg, Aug. 24 - 26, 1978, with the objective of formulating the present standing of research work and of the most important problems in this field. Aspects of physical oceanography
and of coastal engineering were to be considered. Estuarine physics were a suitable topic as both disciplines have many points
in common there.

Eighty-five scientists from sixteen countries convened for two
and a half days to discuss details with much interest. We wish to
thank them as well as the authors for their contributions. Further,
we wish to express our gratitude to the German Research Foundation
(Deutsche Forschungsgemeinschaft), Bonn-Bad Godesberg, to the
Special Research Project 79 "Research in Coastal Waters" (Sonderforschungsbereich 79 "Wasserforschung im Küstenbereich"), and to
the German Hydrographic Institute (Deutsches Hydrographisches
Institut), Hamburg, for sponsoring the meeting and for helping
with the organization.

The authors have agreed to dedicate this book to Prof. Walter
Hansen, who participated in the conference and presided as its
chairman. He has contributed fundamental work and essential impulses
to the discipline of numerical modelling in oceanography.

Hamburg, Hannover, May 1979

> J. Sündermann
> K.-P. Holz

## CONTENTS

Basic Hydrodynamics and Thermodynamics

W. Krauss

Institut für Meereskunde, Kiel


## 1. Introduction

The description of hydrodynamic processes in the sea is based on the
conservation laws for mass (continuity equation), partial mass (diffusion equation for salt), heat (energy equation or equation of heat con-
duction), and momentum (equation of motion). This yields six equations
for the seven field functions $\rho$, S, T, u, v, w and p. The system is
closed by the equation of state.

At the free surface and at the bottom the equations are replaced by
boundary conditions. The proper formulation of these conditions at the
sea surface is essential for any solution, because all driving forces
(besides the tidal forces) act only upon the sea surface and, therefore,
appear in the boundary condition.

Any numerical solution of these equations requires a space- and time-
averaging according to the grid system or the truncation in spectral models. Due to this procedure, small-scale processes appear as fluxes of
momentum, heat or salt in the averaged equations. The parametrization
of these fluxes is one of the major unsolved problems in oceanography.
There is evidence that standard parametrization methods may not be applicable in the sea.

## 2. The set of equations

If the hydrodynamic equations are averaged over scales appropriate to
the grid scale, they take the following form:

(1) $\frac{\partial \vec{v}}{\partial t} + \vec{v} \cdot \nabla \vec{v} + 2\vec{\Omega} \times \vec{v} = - \frac{1}{\rho} \nabla p - \nabla \phi - \nabla_h \phi_G - \frac{1}{\rho} \nabla \cdot \rho \overline{\vec{v}' \vec{v}'}$

(2) $\frac{\partial \rho}{\partial t} + \nabla \cdot \rho \vec{v} = 0$

(3) $\frac{\partial S}{\partial t} + \vec{v} \cdot \nabla S = - \nabla \cdot \overline{\vec{v}' S'}$

(4) $\frac{\partial T}{\partial t} + \mathcal{W} \cdot \nabla T = - \nabla \cdot \overline{\mathcal{W}'T'}$

(5) $\rho = \rho(p,S,T)$

The symbols have the standard meaning, $\phi_G$ is the tidal potential, the primed quantities represent the subscale processes, the overbar means a time <u>and</u> space averaging and the molecular transfer terms have been omitted.

If we apply the same procedure to the boundary conditions at the free surface $z = -\zeta(x,y,t)$, the kinematic and dynamic boundary conditions take the form

(6) $\frac{\partial \zeta}{\partial t} + \mathcal{W}_h \cdot \nabla \zeta + w = - \overline{\mathcal{W}_h' \cdot \nabla \zeta'}$

(7) $(\Pi_I - \Pi_{II}) \cdot \mathcal{W} = 0$

where $\zeta$ is the averaged sea surface. The dynamic boundary condition (7) states that the normal projection of the average stresses at both sides of the sea surface must be equal. Additionally, if $\vec{q}$ represents the flux of heat or salt, we have

(8) $\vec{q}_I = \vec{q}_{II}$

as boundary conditions for (3) and (4). At the bottom $z = H(x,y)$, these fluxes must be zero and (6) takes the form

(9) $\mathcal{W}_h \cdot \nabla H - w = - \overline{\mathcal{W}_h' \cdot \nabla H'}$

The dynamic boundary condition (7) has to be replaced by an equilibrium condition at the bottom. This may either be a slip or non-slip condition depending on physical reasonings.

## 3. Parametrization of the boundary conditions

We may assume the sea surface to be a randomly fluctuating boundary which can be described by a sum of sinusoidal waves with equally distributed phases in the interval $[0,2\pi]$. Then the right-hand term in (6) vanishes and we obtain the kinematic boundary condition in its usual form

(10) $w = - (\frac{\partial \zeta}{\partial t} + \varnothing_h \cdot \nabla \zeta)$    at $z = -\zeta(x,y,t)$

If the same assumption holds for the small-scale bottom topography,(9) reduces to

(11) $w = \varnothing_h \cdot \nabla H$    at   $z = H(x,y)$.

A major problem arises from the dynamic boundary condition (7), which is generally used in the simplified form

(12) $(p \not k - \mu \frac{\partial \varnothing_h}{\partial z})_I = (p \not k - \mu \frac{\partial \varnothing_h}{\partial z})_{II}$

where $\not k$ is the unit vector in the vertical direction, i.e.

(13)   $p = p_L$    and  $\mu \frac{\partial \varnothing_h}{\partial z} = -\vec{\tau}$

($p_L$ is air pressure, $\vec{\tau}$ is wind stress and $\mu$  an eddy viscosity coefficient.)These equations result from two assumptions:

i) the surface waves do not influence the mean pressure field
ii) the surface layer is a turbulent layer in which the small-
    scale processes may be parameterized by an eddy viscosity
    coefficient.

We adopt the second assumption, which then allows to treat (8) similarly as

(14) $\mu_S \frac{\partial S}{\partial z} = S(P-E)$    and   $\mu_T \frac{\partial T}{\partial z} = \frac{Q}{c_p}$

where $\mu_S$, $\mu_T$ are appropriate eddy coefficients, P-E  the difference between precipitation and evaporation and Q is the amount of heat supplied by the atmosphere. Statement i), however, is obviously not true. As known from the theory of radiation stress (M.S. Longuet-Higgins and R.W. Stewart (1964)) the mean pressure is influenced by a wave field yielding effects like wave set up etc. In shallow water this may considerably contribute to the mean sea level. The resulting barotropic current field may not be negligible.

## 4. Parametrization of the fluxes in the interior of the sea

Applying (13) and (14) as boundary conditions we tacitly assume a viscous sublayer at the sea surface. This seems to be justified. The wind-
mixed layer between the sea surface and the Ekman depth is dominated
by strong shear currents and the local structure of turbulence in this
layer may be parameterized by a turbulent viscosity, $\mu$ . Because $\mu$ is
not a property of the fluid its value may vary with depth and may depend on stratification and wind conditions. The question then rises
whether or not a similar assumption is justified for the interior of
the sea.

### 4.1. The standard method of parametrization

The standard parametrization of subscale processes in oceanography is
based on the concept attributed to Boussinesq (1877). He suggested that
the turbulent shear stress due to the fluctuating velocities should be
proportional to mean velocity gradient. According to the strongly non-
isotropic behaviour of mixing processes in extremely large and shallow
areas like the sea the commonly used parametrization of the flux term
in (1) is

$$(15) \quad \nabla \cdot \rho \, \overline{\vec{v}' \vec{v}'} = -\mu \, \frac{\partial^2 \vec{v}}{\partial z^2} - A \, \Delta_h \vec{v}$$

where $\mu$ and A are eddy viscosities representative for the vertical and
horizontal scales.

If we adopt this parametrization - or any more sophisticated one which
relates the fluctuation to the mean field (a summary of recipes is given by Launder & Spalding, 1972) - we have to answer the question: is
there any evidence that the fluctuations observed in the sea are related to the gradients of the mean fields?

### 4.2. The spectral range of unresolved motions

The fluctuating field in the ocean consists of eddies and waves. Geostrophic eddies due to nonlinear effects typically have scales of less
than 100 km and periods of 50 - 300 days (P. Rhines, 1977). Another
class of eddies is produced according to the potential vorticity theo-
rem if (wind-produced) currents interact with bottom topography. Examples
for the Baltic Sea have been given by J. Kielmann (1978). These eddies

have similar horizontal scales, their lifetime, however, depends on the forcing fields (i.e. wind).

Waves cover a wide range in the frequency-wavenumber-domain. Typical examples are Rossby waves, topographic Rossby waves and shelf waves, barotropic and baroclinic tidal waves, inertial waves, internal waves and surface waves. The horizontal scales range from meters to hundreds of kilometers and the periods from hours to weeks. Current measurements have shown that in all parts of the ocean the kinetic energy of the mean currents is considerably less than the energy of the fluctuations.

Parametrization of that part of the fluctuations which cannot properly be resolved in numerical models depends not only on the grid distance and the time resolution. A physical meaningful parametrization depends also on the form of the spectrum.

In Fig. 1 we display a hypothetical spectrum of the wave processes in a stratified sea. The energy should be concentrated in three scales:

   i)  the small scale of the surface waves (sea state) which
       covers approximately the range 1 sec $< \tau <$ 20 sec and 1 cm
       $< \lambda <$ 500 m. Their energy is concentrated in the wind-mixed
       surface layer
   ii)  the mesoscale, occupied by internal and internal inertial
       waves. They cover the entire depth range; the energy is con-
       centrated within $N^{-1} < \tau < f^{-1}$ and 100 m $< \lambda <$ 100 km
  iii)  the large scale which includes the long waves (tides, seiches,
       storm surges), the quasisteady response to meteorological
       forcing and the quasigeostrophic motions.

Standard numerical models are not able to resolve the small-scale and the mesoscale range. The processes of these ranges have to be parame-terized. Furthermore, inertial waves in enclosed basins do not show up in barotropic models because the vertically averaged currents are zero. On the other hand, the inertial waves are the strongest signal in any current spectrum (fig. 2). Their modelling depends critically on the exact knowledge of the wind field. Additionally, the vertical shear of the inertial currents is of considerable importance for the internal waves (see below). These features cannot be properly described by layer-ed models. We therefore propose to <u>parameterize the entire range with periods less than the inertial period</u>.

6

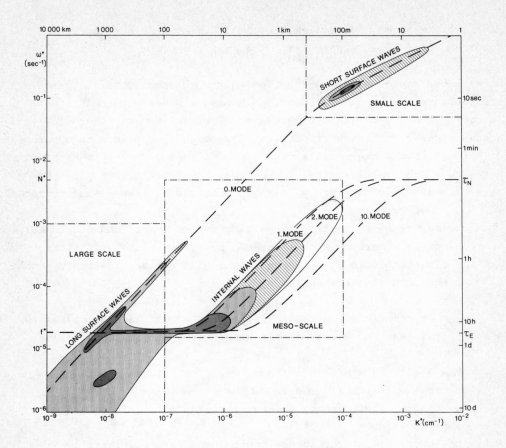

**Fig. 1:** Hypothetical spectrum of
the kinetic energy in a
frequency-wavenumber plane

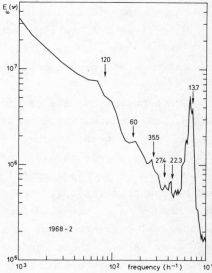

**Fig. 2:** Energy spectrum of the currents
in the Baltic Sea

7

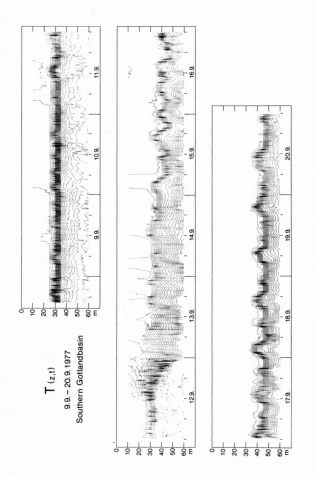

T$_{(z,t)}$

9.9. – 20.9.1977
Southern Gotlandbasin

Fig. 3: Temperature distribution in the Baltic Sea
in September 1977

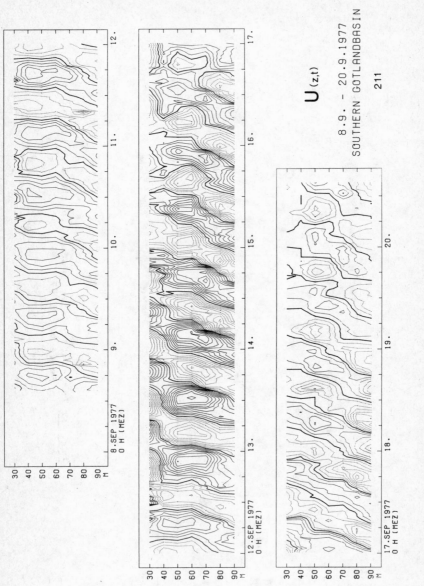

**Fig. 4:** East component (u) of the currents between 30 and 90 m depth (each line corresponds to 5 cm/sec; heavy lines are zero currents)

$U_{(z,t)}$

8.9. – 20.9.1977
SOUTHERN GOTLANDBASIN
211

9

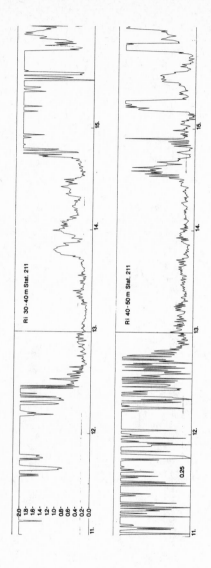

Fig. 5: Richardson numbers for the layers 30-40m and 40-50m

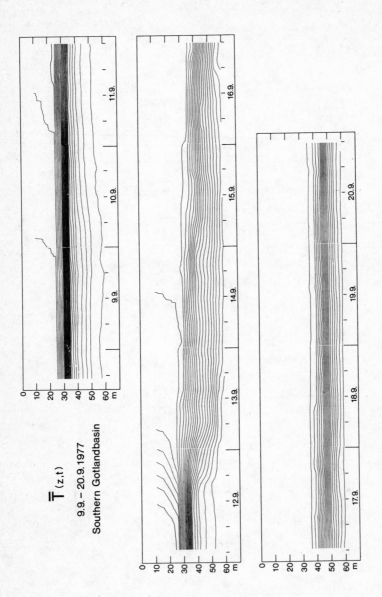

$\overline{T}_{(z,t)}$

9.9. – 20.9.1977

Southern Gotlandbasin

**Fig. 6:** Mean temperature distribution (averaged over one inertial period)

## 5. Physical processes due to inertial waves

The influence of inertial waves on the mean field may be elucidated by
observations from the Baltic Sea. Fig. 3 displays the temperature distribution in the central Baltic (Gotland Basin) during September 1977.
At September 12th a severe storm passed the area and within half a day
the strong thermocline in about 30 m depth was eroded. The mixing started at the top of the thermocline and reduced the temperature in the
mixed surface layer from $15^{\circ}C$ to $12^{\circ}C$. During the afternoon the mixing
penetrated through the entire thermocline. Short internal waves of high
amplitude occurred.

The current field (east component) is shown in fig. 4. It is entirely
dominated by inertial waves. During the passage of the storm (12 September) a system of inertial waves is created at about noon time which
yields a very intensive shear zone between 20 and 30 m (not shown in
fig. 4). This shear zone is further intensified in the afternoon and
reaches maximum values between 30 and 50 m (80 cm $sec^{-1}$/20 m). Due to
this shear the instantaneous Richardson number $\frac{g}{\rho} \frac{d\rho}{dz} \mid [(\frac{du}{dz})^{2} + (\frac{dv}{dz})^{2}]$ is
reduced to values less than 0.25, as shown in fig. 5. Consequently,
short periodic internal waves must break and mixing occurs in the entire
water column between surface and 60 m depth. Thus, mixing is a very rapid process. If the meteorological conditions are such that inertial
waves with intensive shear zones are created, internal waves break and
this process redistributes the heat within a few hours. Mixing below
the Ekman layer is not directly proportional to the wind speed but depends on the inertial waves created by the wind field. In the present
case the inertial waves are enhanced by the passage of a very sharp
meteorological front. The mean temperature distribution is shown in
fig. 6. This example clearly demonstrates that mixing cannot be described
by a proportionality to the gradients of the mean field (Boussinesq).
The energy is directly fed into the inertial waves and these waves yield
a change of the mean potential energy of the fluid.

## 6. Parametrization of the mesoscale range

With respect to inertial waves it is well known that they are directly
produced by meteorological wind fields. The same holds for a wide class
of motions in the entire mesoscale range. The meteorological fields contain enough energy in that frequency and wave number range to produce
internal waves directly due to wind and pressure variations. This is

especially valid in meteorological fronts (W. Krauss, 1978 a,b). Furthermore, variable bottom topography scatters energy into these ranges. We,
therefore, conclude that a parametrization of the mesoscale range is
not possible at present and the Reynolds stresses in equations (1) -
(5) must be retained for a proper description of the slowly varying
part of the spectrum. The inertial waves shown in fig. 4 show horizontal
variations due to bottom topography of 40 cm sec$^{-1}$/20 km, which yields
Reynolds stresses $\frac{\partial u'^2}{\partial x} \approx 4 \cdot 10^{-4}$. These "forcing terms" produce "mean currents" of more than 10 cm/sec within one day.

The closure problem may be solved in the following way:

In order to close the system (1) - (5) we may assume a viscous Ekman
layer, characterized by an eddy viscosity which is gradually reduced to
values of one or less below the Ekman depth (W. Munk, 1966). We then
can compute the fluctuating field due to meteorological forcing in that
frequency wavenumber range and due to bottom topography if the mean conditions can be assumed to be steady for the integration time. The resulting Reynolds stresses may then be used to solve the system (1) - (5).

## References

BOUSSINESQ, J., (1877): Theory de l'écoulement tourbillant. Mem. Pre.
  par div. Sav. 23, Paris
KIELMANN, J., (1978): Mesoscale eddies in the Baltic. Proceedings of the
  XI Conf. Baltic Oceanographers, vol. 2, 729-755
KRAUSS, W., (1978a): The response of a stratified viscous sea to moving
  meteorological fronts and squall lines. Dtsch. hydrogr. Z. 31, 16-30
KRAUSS, W., 1978b): On the energy of the wind stress required to pro-
  duce internal and inertial waves. Dtsch. hydrogr. Z. 31, 31-49
LAUNDER, B.E. & D.B. SPALDING (1972): Mathematical Models of Turbulence.
  Acad. Press. London
LONGUET-HIGGINS, M.S. & R.W. STEWART (1964): Radiation stresses in
  water waves; a physical discussion, with applications. Deep-Sea Res.,
  11, 529-562
MUNK, W. (1966): Abyssal recipes. Deep-Sea Res. 13, 707-730
RHINES, P., (1977): The dynamics of unsteady currents. The Sea. Vol. 6,
  189-361. J. Wiley, New York

Figure Captions

Fig. 1 Hypothetical spectrum of the kinetic energy in a frequency-wavenumber plane

Fig. 2 Energy spectrum of the currents in the Baltic Sea

Fig. 3 Temperature distribution in the Baltic Sea in September 1977

Fig. 4 East component (u) of the currents between 30 and 90 m depth (each line corresponds to 5 cm/sec; heavy lines are zero currents)

Fig. 5 Richardson numbers for the layers 30-40m and 40-50m

Fig. 6 Mean temperature distribution (averaged over one inertial period)

# MATHEMATICAL MODELLING OF TURBULENCE IN ESTUARIES

W. Rodi

Sonderforschungsbereich 80
University of Karlsruhe
Karlsruhe, FRG

## Summary

The paper discusses model assumptions about the turbulent transport terms appearing in the basic equations governing the distribution of velocity, temperature and concentration of salt or other species in estuaries. Fairly crude assumptions can be made about the terms representing the horizontal transport while the vertical transport terms require more refined modelling. The main part of the paper consists of a review of presently available models for the latter; several calculation examples using various models are presented for estuaries and related flows.

## Introduction

The calculation of estuary problems usually involves the solution of equations for continuity, momentum, and scalar quantities such as mass concentration or temperature. In their exact, original three-dimensional form the equations contain terms expressing the transport of momentum, heat, and mass by the turbulent motion, as will be shown in detail below. Due to the appearance of these turbulence terms, the equations do not constitute a closed set and can be solved only when the turbulence terms can either be neglected or determined by introducing suitable approximations. The question therefore arises how important these terms are, and, when they are important , how they can be determined. The present paper deals with this question. It will be shown that, in estuaries, horizontal turbulent transport is relatively unimportant and does not warrant refined modelling; on the other hand vertical turbulent transport is important and requires realistic simulation. The main purpose of the paper is to give a brief review of the presently available mathematical models for simulating the vertical turbulent transport of momentum, heat and mass in estuaries.

In many estuary calculation methods, the originally three-dimensional equations are integrated over one or two cross-sectional directions to yield respectively two- or one-dimensional methods. The integration introduces so-called dispersion terms which are a consequence of the non-uniform distribution of velocity, temperature, or concentration over the direction over which the integration is carried out. It is important to note that these dispersion terms have nothing to do with turbulence and that the turbulence models discussed in this paper are by no means models for estimating these terms.

## Mean flow equations and the problem of closure

The flow and the distribution of scalar quantities in an estuary as sketched in Fig.1 are governed by the following equations :

$$\frac{\partial U}{\partial x} + \frac{\partial V}{\partial y} + \frac{\partial W}{\partial z} = 0 \qquad (1)$$

$$\frac{\partial U}{\partial t} + U\frac{\partial U}{\partial x} + V\frac{\partial U}{\partial y} + W\frac{\partial U}{\partial z} = -\frac{1}{\rho}\frac{\partial P}{\partial x} - \frac{\partial \overline{u^2}}{\partial x} - \frac{\partial \overline{uv}}{\partial y} - \frac{\partial \overline{uw}}{\partial z} \qquad (2)$$

$$\frac{\partial V}{\partial t} + U\frac{\partial V}{\partial x} + V\frac{\partial V}{\partial y} + W\frac{\partial V}{\partial z} = -\frac{1}{\rho}\frac{\partial P}{\partial y} - \frac{\partial \overline{uv}}{\partial x} - \frac{\partial \overline{v^2}}{\partial y} - \frac{\partial \overline{vw}}{\partial z} \qquad (3)$$

$$\frac{\partial W}{\partial t} + U\frac{\partial W}{\partial x} + V\frac{\partial W}{\partial y} + W\frac{\partial W}{\partial z} = -\frac{1}{\rho}\frac{\partial P}{\partial z} - \frac{\partial \overline{uw}}{\partial x} - \frac{\partial \overline{vw}}{\partial y} - \frac{\partial \overline{w^2}}{\partial z} - g\frac{\Delta\rho}{\rho} \qquad (4)$$

$$\frac{\partial c}{\partial t} + U\frac{\partial c}{\partial x} + V\frac{\partial c}{\partial y} + W\frac{\partial c}{\partial z} = -\frac{\partial \overline{uc'}}{\partial x} - \frac{\partial \overline{vc'}}{\partial y} - \frac{\partial \overline{wc'}}{\partial z} \qquad (5)$$

$$\rho = f\{c\} \qquad (6)$$

where (1) is the continuity equation, (2) to (4) are momentum equations in the x-, y-, and z-directions respectively (the Boussinesq approximation has been invoked), (5) is the transport equation for any scalar quantity (c may stand for mass concentration or temperature) and (6) is the equation of state. U,V,W are mean and u,v,w fluctuating (turbulent) velocities in the

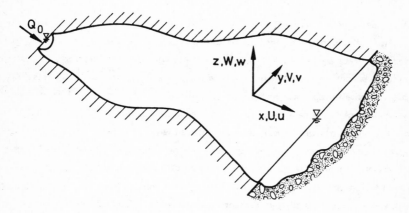

Figure 1 :   Flow configuration

x-, y- and z-directions respectively (see also Fig. 1), P is the static pressure, $\Delta\rho$ the density difference to a reference density, and g the gravitational acceleration. The mean velocities are statistical averages over a time which is long compared with the time scale of the turbulent fluctuations but short compared with that of the mean-flow variations (e.g. tidal period).

The momentum equations (2) to (4) contain correlations between various fluctuating velocity components (the time averaging is indicated by an overbar). These correlations express the transport of mean momentum by the turbulent motion and act on fluid elements like stresses. They are therefore called turbulent or Reynolds stresses. For example the $\overline{uw}$-correlation transports x-momentum in the z-direction (or vice versa) and acts as a shear stress (strictly speaking $-\rho\,\overline{uw}$ is the stress) in the x-direction on a plane normal to z (or vice versa). Similar correlations between the fluctuating velocity components and the fluctuating scalar quantity c' appear in the scalar transport equation (5). These correlations represent the turbulent transport of heat or mass and are therefore turbulent heat or mass fluxes.

Because of the appearance of the various turbulence correlations, equations (1) to (6) do not form a closed set. The main problem in calculating turbulent flows is to determine these correlations so that equations (1) to (6) can be solved. At present, the turbulent transport terms cannot be determined from an exact theory. The exact equations describing the details of the turbulent motion are known (they are the non-averaged counterparts of the above equations), and numerical procedures exist which can solve them in principle; that they cannot be solved in praxis is due to the fact that important processes of the turbulent motion occur at such small scales that they cannot be resolved in a numerical solution with present-day computers. Therefore approximations must be introduced for the turbulence correlations in form of model assumptions, and this can be done only with the aid of experimental information. The assumptions are expressed by equations (algebraic or differential) which allow to determine the turbulent transport terms. These equations constitute a turbulence model; they always contain empirical constants.

One aspect peculiar to estuaries is the large horizontal extent of the flow compared with the extent in the vertical direction. As a consequence, gradients of mean-flow quantities in the horizontal directions are usually small compared with those in the vertical direction, and, since the turbulent transport in a certain direction can be assumed to be closely related to the gradient of the transported quantity in that direction, horizontal turbulent transport is generally much smaller than vertical transport. In particular the horizontal transport terms in the momentum equations are usually small compared with other terms and are therefore mostly neglected. It should however be pointed out that the presence of some diffusive momentum transport is essential in areas with horizontal recirculation since recirculation is not possible without such diffusion in the case of steady flow [1] . In this context it is important to note that some effective diffusion is always present in numerical calculations even when the diffusive transport terms in the equations have been neglected, either unintentionally by numerical diffusion or intentionally by smoothing procedures to procure numerical stability. In contrast to the momentum equations, the horizontal transport terms in the scalar transport equation (5) cannot be neglected; in fact they determine the horizontal distribution of the scalar quantity. However, not only the horizontal gradients of the mean flow quantities are small but also those of the turbulent transport quantities so that refined simulation of these is not warranted, in particular since their effect on the solution cannot be distinguished from the effect of numerical diffusion. Therefore relatively crude models such as a constant turbulent diffusivity assump-

17

tion appear to be sufficient. Here it should be noted that the horizontal diffusivity coefficient, apart from being dependent on numerical diffusion, also accounts for dispersion in the case of depth-integrated calculations and often has to be determined by matching the calculations with observations. The foregoing remarks on horizontal turbulent transport do not apply to the near-field of discharges nor to the details of the flow near the banks, as more refined turbulence models are necessary in these areas. A review on such models can be found in [2].

In contrast to horizontal transport, the vertical turbulent transport of momentum is always important. In depth-integrated methods this transport is accounted for by the bottom friction term; in three-dimensional and two-dimensional width-integrated calculations its distribution over the flow field has to be determined, and a realistic model description is essential. The same applies to the modelling of vertical turbulent heat and mass transport in situations where the estuary is vertically not fully mixed. The following section gives a review of available mathematical models for simulating the vertical turbulent transport of momentum, heat and mass.

## Models for the vertical turbulent transport

The flow in estuaries is of the unsteady channel-flow type in which the vertical momentum equation (4) can normally be reduced to the hydrostatic pressure relation and the flow is predominantly in the x-direction so that it is mainly the x-momentum equation (2) that determines the flow (in width-integrated models this is usually the only momentum equation solved). Consequently, the main problem in modelling vertical turbulent transport is to determine the shear stress $\overline{uw}$ in equation (2) and the heat or mass flux $\overline{wc'}$ in equation (5). This problem is more difficult in estuaries than in ordinary channel flow because of the influence of unsteadiness and buoyancy on turbulence.

Types of turbulence models. Existing models for the turbulent transport can be grouped into two categories: those which employ the turbulent (or eddy) viscosity / diffusivity concept and those which do not. The turbulent viscosity / diffusivity concept assumes that, in analogy to the viscous stresses and fluxes in laminar flow, the turbulent stresses and heat or mass fluxes are proportional to the mean velocity gradients and temperature or concentration gradients respectively. For the transport terms of interest here, this concept yields:

$$\overline{uw} = - v_t \frac{\partial U}{\partial z} \quad , \qquad \overline{wc'} = - D_t \frac{\partial c}{\partial z} \quad , \qquad (7)$$

where $v_t$ and $D_t$ are respectively turbulent (or eddy) viscosity and diffusivity. In contrast to their molecular counterparts, $v_t$ and $D_t$ are not fluid properties but depend strongly on the state of turbulence. The turbulent diffusivity $D_t$ is usually assumed proportional to the turbulent viscosity $v_t$

$$D_t = \frac{v_t}{\sigma_t} \qquad (8)$$

where the proportionality factor $\sigma_t$ is the turbulent Prandtl or Schmidt number. Typical U- and c-profiles are sketched in Fig. 2, from which the behaviour of the gradients in (7) can be inferred.

The assumption of a constant turbulent viscosity / diffusivity in (7) is not adequate for a realistic description of the vertical turbulent transport processes, as it is well known that $v_t$ and $D_t$ have a nearly parabolic distribution with depth in developed open channel flow [3] . Therefore the introduction of relations (7) alone does not specify $\overline{uw}$ and $\overline{wc'}$; the problem is now shifted to determing the distribution of $v_t$. The simplest models use an empirical specification for $v_t$ (e.g. para-

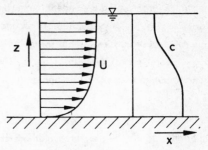

Figure 2 : Typical vertical velocity and concentration profiles in estuaries

bolic distribution) or relate it to the mean-velocity gradient, as is done in the Prandtl mixing length hypothesis. More complex models relate $v_t$ to parameters characterizing the turbulence and determine these parameters from semi-empirical transport equations.

The second group of turbulence models also uses transport equations, but now for the quantities $\overline{uw}$ and $\overline{wc'}$ themselves so that the turbulent viscosity/diffusivity concept, which after all expresses only an assumption, is not introduced. These transport equations involve, of course, also a certain amount of empiricism. The individual models will now be discussed in greater detail.

Prandtl mixing length model. In 1925 Prandtl [4] proposed to relate the eddy viscosity to the mean velocity gradient in the following way

$$v_t = \ell_m^2 \left| \frac{\partial U}{\partial z} \right| , \qquad (9)$$

where $\ell_m$ is the Prandtl mixing length whose distribution over the flow field has to be determined empirically. In channel flow, a mixing length distribution as sketched in Fig. 3 has been found suitable, with the von Kármán constant $\varkappa \approx 0.4$ and $\lambda \approx 0.1$; as indicated in Fig. 3, the mixing length is often reduced somewhat near the free surface in order to account for the restricting effect of this on the size of the turbulence elements.

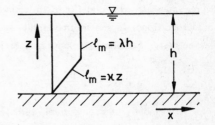

Figure 3 : Mixing length distribution in open channel flow

Buoyancy forces may alter significantly the mixing length distribution. This effect can be accounted for by empirical formulae, in which the influence of buoyancy is characterized by the gradient Richardson number

$$Ri = - \frac{g}{\rho} \frac{\partial \rho / \partial z}{(\partial U / \partial z)^2} . \qquad (10)$$

The formulae originate from studies of stratified atmospheric boundary layers; for stable stratification (Ri > 0 , unstable situations with Ri < 0 are unusual in estuaries) the Monin-Obukhov relation

$$\ell_m / \ell_{m_o} = 1 - \beta_1 \, Ri \qquad (11)$$

is often used, where $\ell_{m_o}$ is the mixing length under unstratified conditions (Ri = 0) and $\beta_1 \approx 7$ (values of $\beta_1$ ranging from 5 to 10 are reported in the literature [5, 6] ). Alternatively, the eddy viscosity $v_t$ can be modified itself by a relation of the form:

$$v_t / v_{to} = 1/(1 + \beta Ri)^\alpha \quad , \qquad (12)$$

where $v_{to}$ is the eddy viscosity under non-buoyant conditions. According to the review study in [7] , the Munk-Anderson [8] relation using $\alpha = 0.5$ and $\beta = 10$ fits best the majority of the experimental data.

When the Prandtl mixing length model is employed, the turbulent heat and mass transport is calculated with the aid of relations (7) and (8), using $\sigma_t \approx 1$. The turbulent Prandtl/Schmidt number $\sigma_t$ is however also altered by buoyancy, and this can be accounted for by another Munk-Anderson formula:

$$\sigma_t / \sigma_{to} = \frac{(1 + \beta_c Ri)^{\alpha_c}}{(1 + \beta Ri)^\alpha} \quad , \qquad (13)$$

with $\alpha$ and $\beta$ as above and $\alpha_c = 1.5$ and $\beta_c = 3.33$.

In Fig. 4, an example is presented of an estuary calculation with salinity intrusion, using a mixing length model with buoyancy modifications of the type (12) and (13). The calculations are for the Hudson river and were obtained by Leonard et al [9] with a 2D width-integrated method. These authors found that the predicted flow behaviour was very sensitive to the value of the constant $\beta_c$ in (13), which they adjusted to match their predictions with field observations. The left of Fig. 4 presents predicted instantaneous stream line patterns and elevations every one-fourth of a complete tidal cycle for a freshwater inflow (at the left) of 100 $m^3/s$ and a sinusoidal tide at the mouth (right) with an amplitude of $\approx$ 1 m. It should be noted that longitudinal and vertical scale are different by a factor of about 1000. The right part of Fig. 4 presents a comparison of predicted and measured cross-section and tidally averaged salinity distribution along the estuary for three different fresh-water flow rates.

The mixing length model has the great advantage of being simple and economical; and Fig.4 indicates that reasonable results can be obtained provided the proper empirical constants are chosen. However, the need to adjust the constants is one of the major drawbacks of the mixing length model and can be traced back to the fact that the model implies the asumption that turbulence is in a state of local equilibrium everywhere, which means that the turbulence produced at a certain point (mainly by interaction of shear stresses and velocity gradients) is dissipated by viscous action at this point at the same rate. Thus the mixing length model does not account for the transport of turbulence (from regions of high to those

20

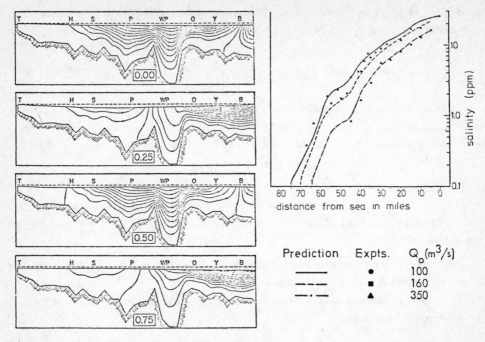

a) Instantaneous stream line patterns and elevation, every one-fourth of a tidal cycle

b) Cross-section and tidally averaged salinity distributions

Figure 4 : Flow and salinity in the Hudson river, width-integrated calculations [9]

of low production), nor for history effects and is therefore not suitable for situations when transport or history effects are important; for example the model grossly underpredicts the turbulent diffusivity for times near slack water, when velocity gradients are low (leading to low $v_t$ and $D_t$ according to (9) ) but in fact turbulence produced at earlier times still persists.

One-equation models. Transport and history effects of turbulence can be accounted for in an eddy viscosity / diffusivity model by solving a transport equations for a suitable parameter characterizing the turbulence. The turbulent kinetic energy, k, is such a parameter as it characterizes the intensity of the fluctuating motion; the following semi-empirical transport equation for k is usually employed in models of this type [2]:

$$\frac{\partial k}{\partial t} + U \frac{\partial k}{\partial x} + W \frac{\partial k}{\partial z} = \frac{\partial}{\partial z} \left( \frac{v_t}{\sigma_k} \frac{\partial k}{\partial z} \right) + v_t \left( \frac{\partial U}{\partial z} \right)^2 + \beta g \frac{v_t}{\sigma_t} \frac{\partial c}{\partial z} - \varepsilon , \quad (14)$$

rate of change | convective transport by the mean motion | diffusive transport by the turbulent motion | shear production | buoyancy production/ destruction | viscous dissipation

where $\beta$ is the volumetric expansion coefficient and $\sigma_t$ an empirical constant. The history effects are accounted for by the rate of change term and the transport effects by the convective

and diffusive transport terms. The distribution of k is influenced further by the shear production term which extracts energy from the mean motion, by viscous dissipation into thermal
energy, and in general also by an interchange between turbulent kinetic energy and potential
energy via the action of buoyancy. The mixing length model implies that the last three terms
on the right hand side balance each other.

In addition to the eddy viscosity / diffusivity assumption introduced already, Eq. (14)
contains only one further model assumption, namely for the diffusion term (gradient diffusion
model). In order to obtain a complete turbulence model, further assumptions are however necessary. The eddy viscosity is, by dimensional analysis, related to k via

$$\nu_t = c'_\mu \sqrt{k} \, L \, , \tag{15}$$

where L is a characteristic length scale of the turbulent motion which also appears in the
following assumption for the dissipation term $\varepsilon$ :

$$\varepsilon = c_D \frac{k^{3/2}}{L} \, . \tag{16}$$

$c'_\mu$ and $c_D$ in (15) and (16) respectively are empirical constants. These two relations together
with the k-equation (14) form a so-called one-equation turbulence model. The distribution of
the length scale L has to be determined empirically in this type of model, and this is done in
a similar way as was described above for the mixing length $\ell_m$. In particular, in non-buoyant
channel flow the length scale can be specified without difficulty and takes a distribution similar to the one sketched in Fig. 3.

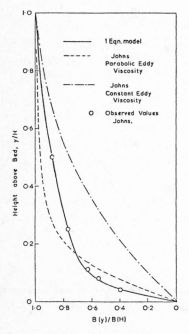

A calculation example for such a flow is given in
Fig. 5 which shows the vertical variation of velocity oscillation amplitude of the tidal flow in the
Humber estuary as predicted by Smith and Takhar
[10] with a one-equation model as described above.
For comparison the figure also includes measurements
as well as predictions obtained with uniform and
parabolic eddy viscosity distributions. It is obvious
from Fig. 5 that the one-equation model, which
accounts for the history effects of turbulence, pre-
dicts the oscillations well while the parabolic eddy
viscosity, which was found suitable for steady channel flows, does not so well.

When buoyancy effects on the turbulen-

Figure 5: Vertical variation of velocity oscillation amplitude of the tidal flow in the
Humber estuary, from [10]

ce are important, the length scale has to be modified and empirical relations similar to the ones described above for the mixing length can be used. Alternatively, the length scale L can be calculated from the following relation :

$$L = \varkappa \, \frac{\psi}{\partial \psi / \partial z} \qquad \text{with} \qquad \psi = \frac{k^{1/2}}{L} \, , \qquad (17)$$

which is a modified form of von Kármán's proposal, the mean velocity gradient in this (i.e., $\psi = \partial U / \partial z$ ) being replaced by a velocity gradient characteristic of the turbulent motion. Since k is calculated from the k-equation (14), which includes a buoyancy term, the effect of buoyancy is accounted for automatically. Relation (17) was developed in the Soviet Union [11] and used there as part of a one-equation model to calculate the flow in various tidal water bodies. Little verification of this formula is however available so far.

Two-equation models. The characteristic length of the turbulent motion at any point in space and time is also subject to transport and history effects; to account for these, models have been developed which use a transport equation also for the length scale L and are therefore more general than the simpler models. The length-scale determining equation does not need to have the length scale L itself as dependent variable; any combination with the turbulent energy k may serve as dependent variable since k is determined separately from Eq. (14). The combination $k^{3/2}/L$ has become most popular so that, according to (16), an equation is solved for the dissipation rate $\varepsilon$ of the turbulent kinetic energy. For channel-type flows this equation reads [12] :

$$\frac{\partial \varepsilon}{\partial t} + U \frac{\partial \varepsilon}{\partial x} + W \frac{\partial \varepsilon}{\partial z} = \frac{\partial}{\partial z} \left( \frac{v_t}{\sigma_\varepsilon} \frac{\partial \varepsilon}{\partial z} \right) + c_{1\varepsilon} \frac{\varepsilon}{k} v_t \left( \frac{\partial U}{\partial z} \right) - c_{2\varepsilon} \frac{\varepsilon^2}{k} \qquad (18)$$

rate   convective      diffusive transport     production minus destruction
of      transport by the    by the turbulent
change   mean motion      motion

where $\sigma_\varepsilon$ , $c_{1\varepsilon}$ and $c_{2\varepsilon}$ are empirical constants. Similar to the k-equation (14), the $\varepsilon$ - equation (18) contains a rate of change term representing history effects, convective and diffusive transport terms, and a source (production) as well as a sink (destruction) term. It should be mentioned however that the $\varepsilon$ -equation is of considerably more empirical nature than the k-equation : all terms on the right hand side of Eq. (18) are based on model assumptions (and therefore contain empirical constants). Together with the k-equation (14) and the eddy-viscosity expression resulting from combining (15) and (16),

$$v_t = c_\mu \frac{k^2}{\varepsilon}$$

the $\varepsilon$ -equation forms the so-called k- $\varepsilon$ -turbulence model, which has been applied successfully to numerous different non-buoyant flows, using the same set of empirical constants [2,12]

A buoyancy-extended version of the k- $\varepsilon$ -model has been developed where, apart from the buoyancy production / destruction term in the k-equation (14), the coefficient $c_\mu$ and the turbulent Prandtl/Schmidt number $\sigma_t$ are functions of some local Richardson number [2, 13 to 15] . These functions were not obtained directly from experiments but were derived by simplifying model forms of the transport equations for the shear stress $\overline{uw}$ and the vertical heat or mass flux $\overline{wc'}$ : in the original transport equations the rate of change and the convective and diffusive transport terms were neglected, but the buoyancy terms were retained; this yielded algebraic expressions for $\overline{uw}$ and $\overline{wc'}$ from which the buoyancy-dependent values of $c_\mu$ and $\sigma_t$ can be extracted. It should further be mentioned here that for horizontal buoyant flows the $\varepsilon$-equation (18) can be used without adding a buoyancy term [14] .

Two-equation models of the type just described have not yet been applied to buoyant estuary situations but were tested for a number of steady flows. Results for one important test case are presented in Fig. 6 which shows the effect of buoyancy on the spreading (left) and the entrainment (right) of a two-dimensional surface jet. In this case a stable stratification is

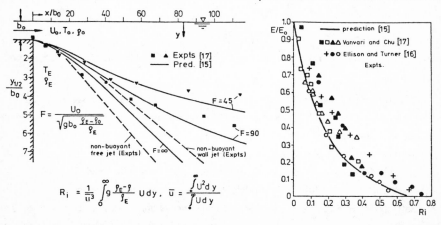

a) Flow configuration and development of half width $y_{1/2}$

b) Entrainment rate ($E_o$ = entrainment rate for non-buoyant jet)

Figure 6 :  Two-dimensional heated surface jet

set up by the surface discharge of lighter fluid which damps the turbulence in the jet and thus reduces its entrainment and spreading. Without tuning the empirical constants this process is described well by the buoyancy-extended k- $\varepsilon$ -model, as can be seen from the comparion with experimental data also shown in Fig. 6. The turbulence damping process is of direct relevance to salt intrusion problems. Fig. 7 presents as a second relevant application example Svensson's [18] width-average calculations of the flow and salinity distribution in the Öresund. The predicted salinity distribution is in general accord with field measurements [18]; in particular the development of a two-layer structure is predicted correctly.

Stress / flux equation models. The turbulence models discussed so far are all based on the eddy viscosity / diffusivity concept. This subsection deals briefly with models which do not use this concept but solve semi-empirical transport equations for the shear stress $\overline{uw}$ and the heat or mass flux $\overline{wc'}$ (for a detailed description see [2]). Such models take direct account of trans-

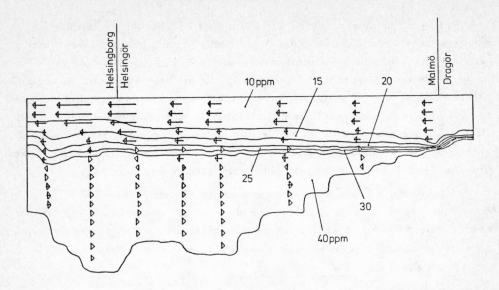

Figure 7 :  Flow and salinity distribution in the Öresund, width-integrated predictions
[18]

port and history effects of $\overline{uw}$ and $\overline{wc'}$ and thus simulate the physical processes most realistically. However, since other turbulence parameters like $k$, $\overline{c'^2}$ and $\varepsilon$ appear in the $\overline{uw}$- and $\overline{wc'}$-
equations, additional equations are necessary to determine these parameters so that this type
of model is altogether rather complex and uneconomical. Simplification of the $\overline{uw}$- and $\overline{wc'}$-
transport equations (e.g. by neglect of the rate of change, convection and diffusion  terms)
leads to the eddy viscosity / diffusivity models discussed earlier, and it seems that these simpler
models are sufficient for estuary calculations. Oscillatory channel flow calculations of Smith
and Takhar [10] with a model using a transport equation for the shear stress $\overline{uw}$ have shown that,
for oscillations of practical relevance, there is only insignificant phase lag between the shear
stress and the velocity gradient so that the eddy viscosity assumption (7) is justified. Since, as
was shown above, buoyancy effects can also be retained when simplifying the transport equations
for $\overline{uw}$ and $\overline{wc'}$ to eddy viscosity / diffusivity expressions, the simpler models based on these
expressions indeed appear sufficiently refined for simulating the vertical turbulent momentum
and heat or mass transport in estuaries.

Concluding remarks

The paper has discussed assumptions about the turbulent transport terms appearing in
the equations governing the distribution of velocity, temperature and species concentration in
estuaries. The horizontal transport was found not to deserve refined simulation because of its
relative unimportance and also because of the relatively small variations of the turbulence terms
in the horizontal directions. The standard practice is to neglect the horizontal turbulent transport terms in the momentum equation and to simulate them in the temperature or concentration
equation with a constant turbulent diffusivity assumption. The vertical turbulent transport on the
other hand does require realistic modelling in 3D and 2D width-integrated calculation proce-

dures. The eddy viscosity / diffusivity concept appears to be a valid approximation for the vertical transport in estuaries, and the usual method of relating the eddy diffusivity to the eddy viscosity via a turbulent Prandtl/Schmidt number is also adequate. The eddy viscosity distribution may be determined to sufficient accuracy with the Prandtl mixing length model in many situations, especially when the empirical constants are tuned to suit the problem. The mixing length model does however neglect transport and history effects of turbulence and is therefore not suitable when such effects are important, as for example during times near slack water. Models employing transport equations for turbulence quantities account for these effects and are therefore more generally applicable and require less, and in many cases no tuning of the constants. Further, buoyancy effects enter naturally in the transport equations and need not be introduced empirically. One-equation models employing a transport equation for the kinetic energy of turbulence appear to be sufficiently complex for non-buoyant estuary situations; they require an empirical length scale specification. Two-equation models which in addition solve an equation for the length scale of turbulence are more generally applicable and are superior particularly in buoyant situations.

## References

1   Flokstra, C., Generation of two-dimensional horizontal secondary currents, Delft Hydraulic Lab. Res. Rep. S 163, part II, (1976).

2   Rodi, W., Turbulence models and their application in hydraulics – A state of the art review, University of Karlsruhe Rep. SFB 80/T/127, (1978) – to appear 1979 as IAHR book publication.

3   Jobson, H.E. and Sayre, W.W., Vertical transfer in open channel flow, Journal of the Hydraulics Division, ASCE, HY3, pp. 703-724, (1970).

4   Prandtl, L., Über die ausgebildete Turbulenz, ZAMM, 5, p. 136, (1925).

5   Arya, S.P.S. and Plate, E.J., Modelling of the stably stratified atmospheric boundary layer, J. Atmos. Sci., 26, pp. 656-665, (1969).

6   Busch, N.E., On the mechanics of atmospheric turbulence, Workshop on Micrometeorology, Am. Met. Soc., pp. 1-65, (1972).

7   Delft Hydraulic Laboratory, Momentum and mass transfer in stratified flows, Report on literature study R 880, (1974).

8   Munk, W.H. and Anderson, E.R., Notes on the theory of the thermocline, J. of Marine Research, Vol. 1, (1948).

9   Leonard. P.B., Vachtsevanos, G.J., Abood, K.A., Unsteady two-dimensional salinity instrusion model for an estuary, Proc. Int. Conf. on Applied Numerical Modelling, University of Southampton, England, (July 1977).

10  Smith, T.J. and Takhar, H.S., The calculation of oscillatory flow in open channels using mean turbulence energy models, Rept. Simon Engineering Labs, University of Manchester, (Nov. 1977).

11  Bobyleva, L.M., Zilitinkevich, S.S. and Laikhtman (Leichtmann) D.L., Turbulent regime in a thermally stratified planetary atmospheric boundary layer, International Colloquium on the Microstructure of the Atmosphere and the Effect of Turbulence on Radiowave Propagation, Moscow, (June 1965).

12    Launder, B.E. and Spalding, D.B., The numerical computation of turbulent flow, Comp. Meth. in Appl. Mech. and Eng., 3, p. 269, (1974).

13    Launder, B.E., On the effect of a gravitational field on the turbulent transport of heat and momentum, J. Fluid Mech., 67, pp. 569-581, (1975).

14    Hossain, M.S. and Rodi, W., Influence of buoyancy on the turbulence intensities in horizontal and vertical jets, in Heat Transfer and Turbulent Buoyant Convection, Studies and Applications for Natural Environment, Building, Engineering Systems, Spalding, D.B. and Afgan, N., eds. Hemisphere Publishing Corp., Washington, D.C., USA, (1977).

15    Hossain, M.S., Mathematische Modellierung von turbulenten Auftriebsströmungen, Ph.D. thesis, University of Karlsruhe, (1979).

16    Ellison, T.H. and Turner, J.S., Turbulent entrainment in stratified flows, J. Fluid Mech., 6, pp. 423-448, (1959).

17    Chu, V.H. and Vanvari, M.R., Experimental study of turbulent stratified shearing flow, J. Hydraulics Div., ASCE, HY6, pp. 691-706, (June 1976).

18    Svensson, U., Flow investigation in the Öresund with a mathematical model (in Swedish), University of Lund, Dept. of Water Resources Engg., Sweden, Rep. No. 3001, (1976).

# FUNDAMENTAL PRINCIPLES FOR NUMERICAL MODELLING

by M.B. Abbott

Reader, International Institute for Hydraulic and
Environmental Engineering, Oude Delft 95, Delft,
The Netherlands and Head, Computational Hydraulic
Centre, Danish Hydraulic Institute, Agern Allé 5,
2970 Hørsholm, Denmark

## 1. The Description Problem: Time, Space and Measures

The first objective of numerical modelling is to describe some aspect of real world
behaviour in terms of numbers and operations on numbers. The second objective is to
make this description in such a way that the value of the information made available
through the modelling process is greater than the cost of the modelling process itself. In a situation where numerical modelling is still winning acceptance, the
value of the information made available must in practice be considerably greater
than the costs. In numerical modelling one is not just looking for "descriptions",
but one is always looking for "best descriptions". In this briefest of introductions,
however, only the barest outline of the first objective can be sketched, even through
the major part of the effort in practical modelling is directed to the second objective.

Because of the special nature of time on the processes of human reasoning (Kant,
1781), it is common to partition the description problem into two parts or processes, namely those describing the "instantaneous" situation and those describing
the transformation from one such situation to another. One asks how to describe
"the state of system at a given time" and how to describe the transformation "from
one state of the system into another state". After the special nature of time in
the processes of human reasoning comes the special nature of space (Kant, 1781).
In mathematics we commonly divide our description between a specification of points
in space and a specification of values of descriptive measures at these points. The
first specification is given in number terms as values of the "independent space
variables" and the second specification is given in terms of the "dependent
variables". It is then usual to subsume the special natures of time and space
specifications under the general nomenclature of "independent variables". A correspondence between dependent variables and independent variables then defines "a
function". The conceptual partitions and their associated number descriptions then
lead to the two principal ways of posing the description problem:

> What are the values of the dependent variables at given values
> of the independent variables? (1)

What are the values of the independent variables at given values
of the dependent variables?                                          (2)

## 2. The Description problem: Discrete and Continuous Descriptions

Another type of partition of the description problem is between discrete descriptions
and continuum descriptions. In the first case the number pairs are said to be countable (or "enumerable" or "denumerable") while in the second case the set of pairs
are said to be "of the order of the continuum". The partition is a fundamental one
in that no countable set, even an infinite one, can be set into a one-to-one correspondence with the continuum (as was presaged by the ancient Greeks through
their discovery of the irrationality of $\sqrt{2}$ and $\sqrt{3}$). However, there exist many (and
in principal infinitely many) ways of otherwise linking these descriptions conceptually.

The prototype of the countable description is the sequence of number-pairs, e.g.

$$(f_0, x_0), (f_1, x_1), (f_2, x_2), \ldots, (f_j, x_j), \ldots (f_J, x_J) \qquad (3)$$

while the prototype of the continuum description is the polynomial

$$f = a_0 + a_1 x + a_2 x^2 + \ldots + a_k x^k + \ldots + a_K x^K \qquad (4)$$

The function (3) defines a set of points in the plane with coordinates f and x while
the function (4) defines a line in this same plane. A special formulation of the
description problem asks: how is a description with the set of points $f_j = f(x_j)$ or
$x = x(f_j)$, related to a description with the line $f = f(x)$ or $x = x(f)$? There are
again, many possible solutions. If $J = K$ then one can construct the polynomial (4)
such that it passes through all the $\{f_j, x_j\}$ of (3). One can, more generally, use
other basis vectors than the $\{x^k\}$ of (4), such as those of orthogonal polynomials,
and one can use other measures of correspondence between the descriptions, such as
least squares approximination.

The modelling process not only necessitates the description of an "instantaneous"
state but it also necessitates the description of a state transformation. For the
last purpose it commonly resorts to a "point description that holds for all points"
in continuum descriptions , and this is usually expressed through a differential
equation. In the discrete description this becomes an "interval description that
applies to all intervals", realised through a finite difference or finite element
formulation. When the approximation is posed in the form (1) we speak of a Dependent
Variable or Direct formulation while when it is posed in the form (2) we speak of
an Independent Variable or Inverse formulation. Thus, in terms of "finite differences"
the differential equation

$$\frac{\partial f}{\partial t} + \frac{\partial f}{\partial x} = 0 \qquad (5)$$

may be approximated by the Direct scheme

$$\frac{f_j^{n+1} - f_j^n}{t^{n+1} - t^n} + \frac{f_j^n - f_{j-1}^n}{x_j - x_{j-1}} = 0 \tag{6}$$

where n is a time address and j a space address and n is no function of j or it may be approximated by the Inverse scheme

$$x_j^{n+1} = x_j^n + (t^{n+1} - t^n) \tag{7}$$

where n is again a time address but now j is an address of the measure f. One of the ways of expressing the modelling relation between (5), (6) and (7) is through the use of the Taylor's series form of (4):

$$f(x) = f(0) + f^{(1)}(0).x + f^{(2)}(0)\frac{x^2}{2!} + \ldots f^{(J)}(0)\frac{x^J}{J!}$$

where $f^{(j)} = \dfrac{d^j f}{dx^j}$ .

Then if (6), for example, provides solutions of polynomials (4) it can be written as

$$0 = \frac{f_j^{n+1} - f_j^n}{t^{n+1} - t^n} + \frac{f_j^n - f_{j-1}^n}{x_j - x_{j-1}} = \frac{\partial f}{\partial t} - \frac{\partial f}{\partial x} + \left[ - \frac{\partial^2 f}{\partial x^2}\frac{(x_j - x_{j-1})}{2!} + \ldots \right.$$

$$\left. + \frac{\partial^2 f}{\partial t^2}\frac{(t^{n+1} - t^n)}{2!} - \ldots \right]$$

The bracketed term on the right is a measure of the difference between (6) and (5) and is called the truncation error of (6) relative to (5).

Now if (6) were used to construct a set $\{f_j^{n+1}\}$ from a set $\{f_j^n\}$ and supplied with suitable boundary data then we could generate an "approximate solution" of (5) using (6). As the $(t^{n+1} - t^n) \to 0$ and the $(x_j - x_{j-1}) \to 0$, by using more and more descriptive numbers, then the truncation error would tend to zero, in this case regardless of how (in which ratios, for example) these increments $\to 0$. We then say that (6) is (in this case unconditionally) consistent with (5). In the limit of this process the truncation error disappears. However, the process remains a countable process expressible in terms of rational numbers (e.g. fractions of the original increments) and cannot provide the set of all numbers, rational and irational, that is comprehended by the set of the power of the continuum upon which (5) is defined. Thus the solution of (6) does not necessarily converge to the solution of (5) in the above limit. In the above example it will converge if $(f^{n+1} - t^n) \le (x_j - x_{j-1})$ while if $(t^{n+1} - t^n) > (x_j - x_{j-1})$ it will diverge, and indeed in the limit the difference between the solutions will be unbounded.

This difference in behaviour, which derives from the non-equivalence of countable sets and sets of the power of the continuum, can be described in terms of the stability of the finite scheme. If the solution remains bounded in the above limit for all operands then the scheme is said to be stable, while otherwise it is said to be unstable. Stability is traditionally investigated by using a further representation of the solution of the discrete scheme. For this representation a basis of orthonormal functions $\{\mathcal{V}_l\}$, so functions that satisfy $(\mathcal{V}_l, \mathcal{V}_m) = \delta_{1m}$, is employed. Thus for example, the functions

$$\{e^{ilx}/\sqrt{2\pi}, l = 0, 1, 2, \ldots\} = \{\mathcal{V}_l\}$$

provide

$$(\mathcal{V}_l, \mathcal{V}_m) = \int_0^{2\pi} \left(\frac{e^{ilx}}{\sqrt{2\pi}} \cdot \frac{e^{-imx}}{\sqrt{2\pi}}\right)dx = \begin{array}{l} 1 \quad \text{when} \quad l = m \\ 0 \quad \text{when} \quad l \neq m \end{array} = \delta_{1m}$$

Then it may be shown that a series

$$S_L = \sum_{l=0}^{L} \xi_l \mathcal{V}_l$$

differs from any function f of a particularly broad class of functions by the smallest amount in the sense of least squares, i.e.

$$\int_R (f - S_L)^2 dx \quad \text{is a minimum} ,$$

when the $\xi_l$ are chosen such that the $\xi_l = (f, \mathcal{V}_l)$. The $\{\xi_l\}$ are then called Fourier Coefficients and the series $S_L$ is called a Fourier Series. In terms of the set (3) viewed directly, $f_j = f(x_j)$, and, for convenience, with $x_j + x_{j-1} = \Delta x$ for all j, the solution of (6) can be written as

$$f_j^n = \sum_k \xi_k^n \exp (i2\pi kj\Delta x/2l) \tag{8}$$

where the k are now dimensionless wavenumbers.

The linearity of the assumed solution enables one to follow the behaviour of one component as typical of all others. Substituting the typical kth component of (8) in (6) provides

$$\xi_k^{n+1} = (1 + \frac{\Delta t}{\Delta x} (\exp (-i2\pi k\Delta x/2l) - 1))\xi_k^n \tag{9}$$

$$= A\xi_k^n$$

so that the kth Fourier Coefficient $\xi_k$ is amplified by the amount A at each time step $(t^{n+1} - t^n) = \Delta t$. Plotting (9) out in the Argand diagram of complex numbers shows indeed that $|A| \leqslant 1$ only so long as $\Delta t/\Delta x \leqslant 1$, the result already mentioned above.

Of course it may be questioned whether the approximation of any f with an $S_L$ is sufficient or of such a nature that we can rely on the result of this analysis. In fact, the basis $\{e^{ilx}/\sqrt{2\pi}\}$, involving complex numbers, is not generally satisfactory in this respect, but it appears that for the type of problems studied in hydraulics, and especially for hydrodynamic studies, it suffices for most practical purposes.

Another question that naturally arises in discrete modelling is whether there is "nothing between stability and instability" - are there only the two possibilities of convergence to the solution of the continuum solution and instability? For linear equations with linear coefficients, such as (5), (6), (7), it may be shown that there is indeed "nothing in between". However, for non-linear equations there are cases where the truncation error does not disappear in the limit $t^{n+1} - t^n \to 0$, $x_j - x_{j-1} \to 0$. The study of these cases continues to provide an interesting area of research in computational hydraulics (e.g. Abbott, 1979).

References

Kant, I. Kritik der reinen Vernunft, Hartknoch, Riga, 1781, (Insel, Wiesbaden, 1956).

Abbott, M.B., Computational Hydraulics; Elements of the Theory of the Free-Surface Flows, Pitman, London, 1979.

Symposium "Mathematical modelling of estuarine physics"
Hamburg, August 24-26, 1978

# APPLICATION OF FINITE-DIFFERENCE METHODS TO ESTUARY PROBLEMS

dr. C.B. Vreugdenhil
Delft Hydraulics Laboratory
Delft, The Netherlands

## Abstract

A variety of mathematical-physical problems is met in the study of estuarine physics.
There is no "universal" method to solve all of them and ad-hoc considerations play a
part in the choice and elaboration of numerical methods. This does not preclude, however, that some basic issues must be considered for each method. They can be combined
under the heading of accuracy. Such considerations are certainly not restricted to
finite-difference methods.

Some examples are given, concerning:
- wave propagation in hyperbolic systems
- representation of flow patterns
- parabolic systems
- treatment of boundaries, particularly fixed walls

Some current research projects will be used to illustrate the above-mentioned subjects:
- two-layer flow in two horizontal dimensions, to be used, e.g., for salt intrusion
  in an estuary
- a vertically two-dimensional model of tidal flow in an estuary, either or not taking
  density stratification into account
- a vertically two-dimensional quasi-steady model of suspended sediment transport, to
  be used, e.g., for the study of sedimentation of dredged trenches in a tidal region

# 1 Introduction

Application of finite-difference methods to estuary problems is now so wide-spread
that a detailed review of such methods is not given here. Some useful references in
this respect are TRACOR (1971), Mahmood and Yevjevich (1975), Vreugdenhil and Voogt
(1975). However, experience shows that quite a few unsolved or incompletely solved problems remain, concerning both physical and numerical aspects. The purpose of the present
paper is to indicate some of the numerical problems. The main message is that one should
be very careful to ascertain that a finite-difference method, or, for that matter, any
numerical method, really produces what it is supposed to produce and does not obscure
the physics one is interested in. This is not to discourage use of numerical methods,
but rather to encourage their critical use, taking into account the peculiarities of
each application.

## 2  Wave propagation in hyperbolic systems

Most of the mathematical models used to study estuary problems are of a parabolic nature; yet hyperbolic systems, with particular reference to tidal-wave propagation in
one or two dimensions, have been studied more thoroughly. The numerical representation
of wave propagation is rather well-known, as expressed by amplitude and phase properties
of finite-difference methods (Kreiss and Oliger, 1973, Leendertse, 1967). Some aspects
seem to be less well-known and they are discussed below.

Amplitude and phase errors are defined as follows. The damping rate d per wave period
T is (for linear wave propagation):

$$d = \frac{\text{amplitude numerical wave at } t = T}{\text{amplitude analytical wave at } t = T} \qquad (2.1)$$

starting from a common initial condition. The relative velocity of propagation

$$c_r = \frac{1}{2\pi} \text{ (numerical phase shift at } t = T) \qquad (2.2)$$

The two parameters are related to the amplification factor $\rho$ per time step for a particular difference method by

$$d = |\rho|^{2\pi/\sigma\xi} \qquad (2.3)$$

$$c_r = -\arg(\rho)/(\sigma\xi) \qquad (2.4)$$

where $\xi = k\Delta x = 2\pi\, N^{-1}$

$\sigma = c\Delta t/\Delta x$ Courant number

$k$ = wave number

$\Delta x$ = mesh width

$N$ = number of grid points per wave length

$c$ = velocity of propagation of waves

$\Delta t$ = time step

If a certain accuracy $\epsilon$ is required, one should have

$$|1-d| < \epsilon \qquad \text{and} \qquad |1-c_r| < \epsilon \qquad (2.5)$$

For two typical finite-difference methods, application of these definitions leads to
the following requirements (approximation for small $\xi$ or large N, i.e. for long waves)

$$(\frac{3}{2\pi^2}\frac{\epsilon})^{\frac{1}{2}} \; N \; > \; (1-\sigma^2)^{\frac{1}{2}} \qquad \text{leap-frog scheme} \qquad\qquad\qquad (2.6)$$

$$(\frac{3}{2\pi^2}\frac{\epsilon})^{\frac{1}{2}} \; N \; > \; (1+\tfrac{1}{2}\sigma^2)^{\frac{1}{2}} \qquad \text{Crank-Nicholson scheme} \qquad\qquad (2.7)$$

Both requirements come from the phase-error, as the two schemes do not have any amplitude error. The Crank-Nicholson scheme is an implicit one, so that the Courant number
$\sigma$ may exceed unity (unlike the leap-frog scheme), but it is seen that accuracy is
maintained only by increasing N. An indication of the relative efficiency of the two
methods can be obtained by determining the amount of work, needed to cover a physical
area of (1 wavelength)$^2$ during 1 wave period, covered by $N^3/\sigma$ grid points. Although
the expressions (2.6) and (2.7) formally apply only to wave propagation along a coordinate axis, they are supposed to be indicative for the general two-dimensional behaviour. Not counting the number of operations required to advance one grid point over
one time step (which differs for the two methods by a factor of, say, 5–10), the amount
of work $W = N^3/\sigma$ is plotted in Figure 1 as a function of the Courant number $\sigma$. For the
leap-frog method, the "ideal" value $\sigma = 1$ is not attainable in practice, as $\sigma$ is

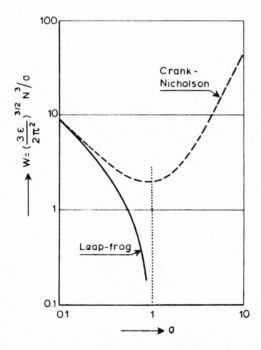

Figure 1   Relative amount of work W for accuracy $\epsilon$

not constant throughout the model. Normally, one will try to operate with
$\sigma \approx 0.7$ as a mean value, for example. For the Crank-Nicholson, the most striking
feature is that the amount of work is minimal for $\sigma = 1$. At higher values of
the Courant number, the amount of work required for the accuracy $\varepsilon$, increases
quite rapidly. The conclusion must be that, at least from an economical point of view,
implicit methods do not "pay" for this type of wave propagation problems.

There may be additional considerations leading to the use of implicit methods. One is
the representation of geometry (boundaries, bathymetry), which may require a smaller
mesh width than follows from accuracy of wave propagation, discussed above (the question
whether such a detailed representation of geometry is really necessary is not elabo-
rated here). Figure 2 shows what happens to the amount of work if the number of points
per wave length N must exceed a certain number $n_o$ due to geometrical reasons. For

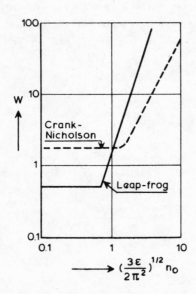

Figure 2   Amount of work under geometrical constraint

small $n_o$ this does not give any restriction (wave propagation is the determining factor). For large $n_o$ it is seen that the Crank-Nicholson method becomes relatively less
expensive, due to the possibility to use a Courant number exceeding unity in that case.

Another situation where the question of economical use of explicit or implicit methods
comes up is the case where wave speeds of different magnitude occur. An example is
two-layer density stratified flow in estuaries, where the celerities of surface and
internal waves differ by a factor of about 10. There are some possibilities of choosing

finite-difference methods (Figure 3);

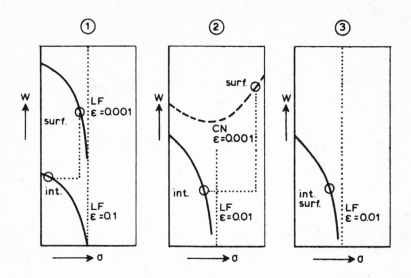

Figure 3   Three possibilities for computation of surface and internal waves

i      Both wave types are treated by the leap-frog method on a common grid. The surface
       waves then limit the Courant number $\sigma_s$ to about 0.7, so that the internal-wave
       Courant number $\sigma_i$ will be about 0.07. The number of grid points $N_s$ per surface
       wave length will be about 10 times the corresponding number $N_i$ for internal waves,
       which means that the surface waves are computed to a much higher accuracy than
       the internal waves.

ii     Internal waves are computed explicitly at $\sigma_i \approx 0.7$; surface waves implicitly at
       $\sigma_s \approx 7$. Still, $N_s \approx 10 \, N_i$ so that some unbalance in accuracy remains, although
       this method is more efficient than the preceding one (Parot, 1976).

iii    Both wave types are done explicitly but using different grids, such that $N_s \approx N_i$
       and $\sigma_s \approx \sigma_i$. The accuracy for surface waves is now about the same as that for
       internal waves (Vreugdenhil, 1978).

In the latter two cases, it is necessary to (approximately) uncouple the equations for
surface and internal waves. Indeed it turns out to be possible to write the equations
in a "weak interaction" form (Vreugdenhil, 1978) in which the interaction terms are
relatively small so that they may be approximated at a somewhat lower accuracy, if
necessary. An example of a computation on this basis (however using method i above)
is shown in Figure 4 where surface (left) and interface (right) positions are shown
for a schematic estuary during the third tidal cycle. The left and right boundaries
are open; they are treated as non-reflecting boundaries to outgoing waves. Incoming
waves are specified as travelling waves from infinity.

38

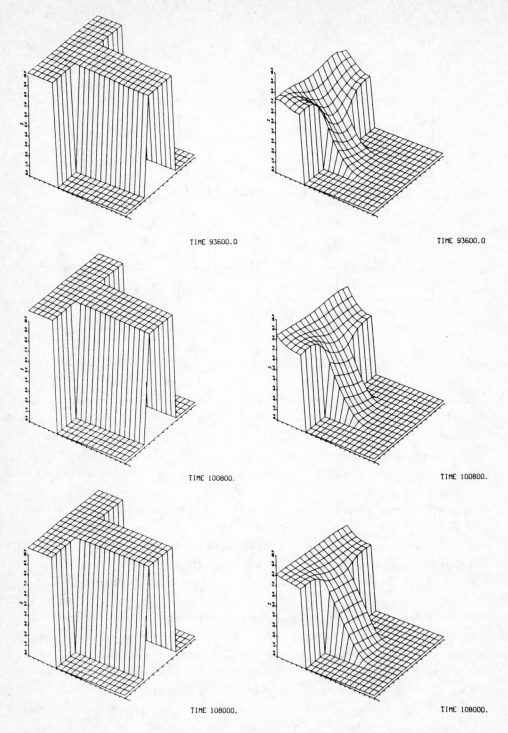

TIME 93600.0                                                 TIME 93600.0

TIME 100800.                                                 TIME 100800.

TIME 108000.                                                 TIME 108000.

Figure 4  Example of  two-layer flow; surface positions (left) and interface
          positions (right) at two-hour intervals

In the example of Figure 4 a numerical difficulty was met: oscillations with a short period (2Δt) but a rather large wave length occurred which were not sensitive to any spatial smoothing. They could, however, be suppressed very well by averaging in time every 40th time step (Figure 5). The occurrence of long-wave parasitic modes seems to

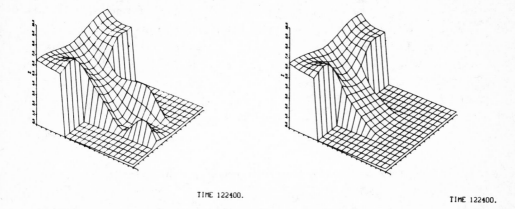

TIME 122400.

TIME 122400.

Figure 5   Long wave oscillations (left) suppressed by time averaging (right)

be unknown. In the present case it can be explained by analysing the conditions under which the amplification factor ρ may be − 1 for the leap-frog scheme. If the differential equations are written in the quasi-linear form (disregarding Coriolis and friction effects)

$$v_t + A\,v_x + B\,v_y = 0 \qquad\qquad (2.8)$$

it turns out that ρ = − 1 if

$$\text{determinant } (A \sin \xi + B \sin \eta) = 0 \qquad\qquad (2.9)$$

with $\xi = k_x \Delta x$ and $\eta = k_y \Delta y$. One condition in which Equation (2.9) is satisfied is ($\xi$, $\eta$ small)

$$k_x\,u + k_y\,v = 0 \qquad\qquad (2.10)$$

which describes a (possibly long) wave, propagating normal to the flow vector (u, v). This is approximately what can be seen in Figure 5.

## 3  Flow patterns

For the estimation of the accuracy with which flow patterns are reproduced there does
not seem to exist any systematic method, although it may be related to slowly propagating wave modes, comparable to Rossby waves in meteorology (e.g. Elvius and Sundström,
1973). Moreover, it is known from experience that flow patterns (and particularly mean
flows) are more sensitive than water level variations. Analysis must therefore be of
a more qualitative character. One important aspect is numerical diffusion. This is a
well-known drawback of low-order finite difference methods, but it must be realised
that higher order methods also give rise to numerical diffusion if applied to non-
linear equations. For example, solving the transport equation

$$u_t + \{f(u)\}_x = 0 \tag{3.1}$$

by means of the leap-frog method, the truncation error is found from

$$u_t + f_x + \tfrac{1}{3} \Delta t^2 \, u_{ttt} + \tfrac{1}{3} \Delta x^2 \, f_{xxx} + \ldots = 0 \tag{3.2}$$

For a linear equation, with $f = cu$ and $c$ = constant, the leading term of the truncation
error is a third derivative, which does not have a diffusive effect (of course higher
even-order terms do!). However, for a non-linear equation

$$f_{xxx} = f''' \, u_x^3 + 3 \, f'' \, u_x \, u_{xx} + f' \, u_{xxx} \tag{3.3}$$

and it is seen that the middle term does introduce diffusion, with a coefficient proportional to $\Delta x^2$ and $u_x$, so that it will be most important in regions of strong gradients. Similar effects are found, e.g., in some of the Arakawa schemes (Arakawa,
1966).

An example of the influence of numerical diffusion (or viscosity) is given in Figure 6
where a steady flow pattern in a harbour basin along a river is shown, using two
different values of the numerical viscosity coefficient $\varepsilon$ (which was introduced by
means of a smoothing process). It is seen that the magnitude of the circulation is
considerably influenced by the difference in viscosity. The actual physical magnitude
of the coefficient of viscosity seems to be unknown as yet, so that prediction of this
type of flows remains uncertain.

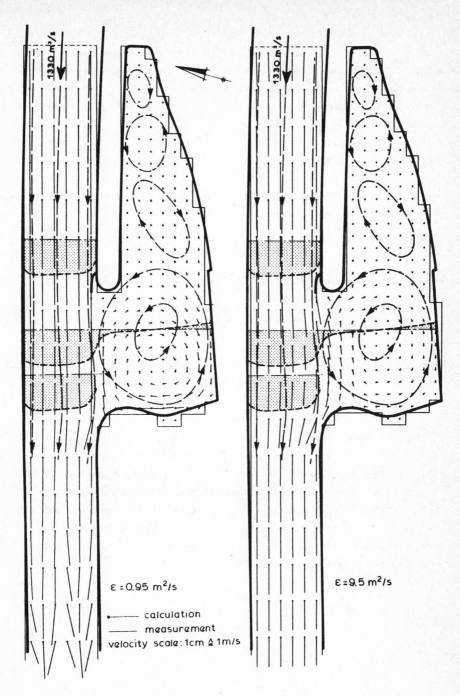

Figure 6  Example of circulatory flow with two different values of viscosity
coefficient

In a sense, the prediction of such coefficients is a matter of subgrid-scale modelling. Those effects due to depth averaging must be taken into account in a parametrical form; in addition, it has sometimes been suggested to "model" the contributions due to averaging along grid intervals in a similar way (a.o. Vreugdenhil and Voogt, 1975). However, this does not appear to be very useful due to the following reason. If one has a conservation law

$$h_t + f_x + g_y = 0 \tag{3.4}$$

a finite difference approximation can be generated by integrating over a grid square:

$$\frac{\partial}{\partial t} \iint h \, dxdy + \oint (f, g) \cdot \underline{n} \, ds = 0 \tag{3.5}$$

which is still exact. Approximating the integrals gives, e.g., central differences:

$$\frac{\partial h_o}{\partial t} + D_{0x} f + D_{0y} g = - \frac{\partial}{\partial t} \iint (h - h_o) \, dxdy - \int_{-\Delta y}^{\Delta y} \{f(x + \Delta x, y) - f(x + \Delta x, 0)\} \, dy \, \ldots \tag{3.6}$$

where $h_o$ is the value of h in the grid centre. The right-hand side is an exact representation of the truncation error, and it describes the "subgrid dispersion". Formally it is equivalent to the truncation error obtained by the more usual development into Taylor-series. Two observations can now be made:

i   If one would model the fluxes across grid sides, one should also include the surface integral in (3.6) which is of the same order.

ii  From the equivalence with the truncation error by Taylor series, it follows that the right-hand side is important only if the variables do not vary smoothly; however, exactly in that case it is difficult to see how one could model the truncation-error terms in such a way that the error would be reduced.

Therefore, this does not seem to be way to take into account those effects which were discarded in the original discretization.

## 4 Parabolic systems

In principle, the accuracy of numerical methods for parabolic systems could also be
treated by Fourier methods, as shown in Section 2 for hyperbolic ones. Apart from one-
dimensional models, this does not seem to have been done to any extent. Moreover, an
additional difficulty arises in determining the relevant wave length or frequency to
be used. As a first attempt for vertically two-dimensional models one could consider
the following simplified formulation:

$$u_t + g \, \xi_x - \varepsilon \, u_{zz} = 0 \tag{4.1}$$

$$\xi_t + \frac{\partial}{\partial x} \int_0^h u \, dz = 0 \tag{4.2}$$

$$z = 0 \quad \rightarrow \quad u = 0 \tag{4.3}$$

$$z = h \quad \rightarrow \quad \partial u / \partial z = 0 \tag{4.4}$$

where  u = horizontal (longitudinal) velocity
  $\xi$ = surface elevation
  $\varepsilon$ = eddy viscosity
  h = mean depth
  z = vertical coordinate
  x = longitudinal coordinate

Such systems, in a more complete form, have been studied by, amongst others, Blumberg
(1977) and Perrels and Karelse (1977). Moreover, hydrostatic three-dimensional models
can be expected to have similar properties. In order to characterize the solutions, a
travelling wave is sought:

$$u = U(z) \, \exp \left\{ i\sigma \left( t - \frac{x}{c} \right) - x/L \right\} \tag{4.5}$$

where  c = velocity of propagation
  L = damping length
  $\sigma$ = frequency
  U = vertical velocity profile

Inserting Equation (4.5) into (4.1)...(4.4), assuming constant coefficients, allows
c and L to be computed as functions of the parameter $h(\sigma/2\varepsilon)^{\frac{1}{2}}$ which is related to
the time of vertical mixing compared to the wave period. The result is shown in
Figure 7 and it is not very obvious in that the velocity of propagation can be quite
different from $(gh)^{\frac{1}{2}}$, the value which would follow from a one-dimensional analysis.
The significance of these results has to be evaluated in more detail, and similar
analyses must be made for the numerical systems used.

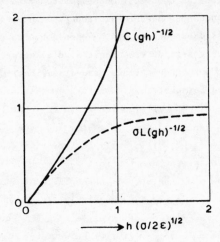

Figure 7   Behaviour of velocity of propagation c and damping length L for
vertically 2-d tidal model

## 5 Boundary conditions

A comprehensive discussion of boundary conditions for various types of mathematical estuary models is outside the scope of this paper, although it is certainly one of the more important problems. A theory of any completeness is not available, but very useful information can be obtained from Kreiss and Oliger, 1973, Sundström, 1977, Garrett and Greenberg, 1977, Engquist and Majda, 1977. Here, the discussion is restricted to the treatment of wall layers in turbulent flow. Near a fixed bottom, the eddy viscosity goes (almost) to zero, which leads to logarithmic velocity profiles ("law of the wall") and exponential profiles for sediment concentration. Normally, such layers cannot be resolved sufficiently well by finite difference methods. Yet, the effect must be taken into account in some way to get correct velocity and concentration profiles farther from the bottom. Depending on the amount of interest in the wall layers themselves, two methods are available. In boundary-layer calculations it is customary to fit the law-of-the-wall profile to a grid point near, but not on, the bottom. This is based on a local equilibrium between the pressure gradient along the wall and the shear-stress gradient normal to it. In unsteady and non-uniform flow the same method can be applied because the additional terms in the momentum equation vanish as $z \to z_o$ (which is the level of zero mean velocity):

$$u_t \quad + uu_x \quad + wu_z \quad + g\,\xi_x \quad - (\varepsilon\,u_z)_z \; = 0$$

$$\ln \frac{z}{z_o} \qquad (\ln \frac{z}{z_o})^2 \qquad \ln \frac{z}{z_o} - \frac{z-z_o}{z} \qquad O(1) \qquad O(1) \qquad \text{if } u \sim \ln \frac{z}{z_o} \tag{5.1}$$

An example of tidal flow obtained this way is shown in Figure 8, taken from Perrels and Karelse, 1977. For further details and extension to density-stratified flow, reference is made to the original publication.

If, on the other hand, one is interested in the wall region itself, the preceding method may be less satisfactory. This is the case, e.g., in suspended sediment transport, where an important fraction of the transport may occur in a thin layer near the bottom. In this case, a coordinate transformation may be used, which stretches the bottom layer in such a way that it can be represented numerically. The steady-state transport equation for suspended sediment

$$u\,c_x + (w + w_s)\,c_z - (\varepsilon_s\,c_z)_z = 0 \tag{5.2}$$

can be transformed by

$$z' = \int_{z_o}^{z} \frac{w_s}{\varepsilon_s}\,dz \tag{5.3}$$

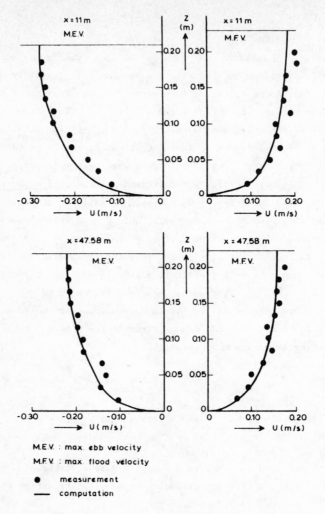

Figure 8  Computed and measured velocity profiles in tidal flume

into the equation

$$\frac{u \, \varepsilon_s}{w_s^2} c_x + \{ \frac{u \, \varepsilon_s}{w_s^2} \frac{\partial z'}{\partial x} + \frac{w + w_s}{w_s} \} \, c_{z'} - c_{z'z'} = 0 \qquad (5.4)$$

in which the latter term is now in a simple form for numerical treatment. Other trans-
formations with the same effect can be envisaged. Figure 9 shows an application of
quasi-steady sediment-transport for a trench, dredged across the Western Scheldt
estuary. The tidal flow is schematized into some blocks giving the correct total transport. Although a quite crude numerical method is used, transport rates and corresponding

rates of sedimentation appear to be very reasonable. Further details can be found in the original reference (Kerssens, van Rijn, 1977).

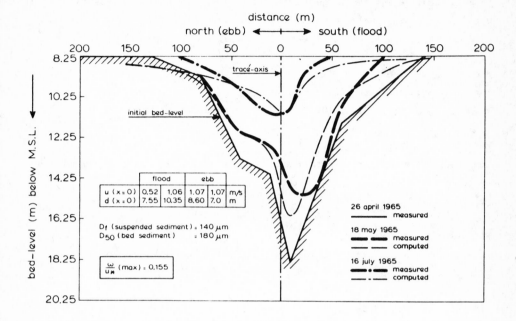

Figure 9   Development of a dredged trench computed with quasi-steady suspended
           sediment model

## 6   Conclusion

The above examples are a rather small and personal choice from the aspects of finite-difference methods for estuary problems. It has been indicated that, although we do have some methods of analysing the numerical problems, quite a few difficulties are left. For any particular application one has to make reasonably sure, using available methods and sound engineering judgement, that the results are sufficiently reliable for the purpose.

# 7 References

1   ARAKAWA, A.,
    Computational design for long-term numerical integration of the equations of fluid
    motion: two-dimensional incompressible flow, Part I,
    J. Comp. Physics 1 (1966), 119-143

2   BLUMBERG, A.F.,
    Numerical model of estuarine circulation,
    Proc. ASCE, J. Hydr. Div., 103, HY3 (1977), 295-310

3   ELVIUS, T., and SUNDSTRÖM, A.,
    Computationally efficient schemes and boundary conditions for a fine-mesh model
    based on the shallow-water equations,
    Tellus 25 (1973) 2, 132-156

4   ENGQUIST, B., and MAJDA, A.,
    Absorbing boundary conditions for the numerical solution of waves,
    Math. Comp. 31, 139 (1977), 629-651

5   GARRETT, C., and GREENBERG, D.,
    Predicting changes in tidal regime: the open boundary problem,
    J. Phys. Oceanography 7 (1977), 171-181

6   KERSSENS, P.J.M., and Rijn, L.C. van,
    Model for non-steady suspended sediment transport,
    IAHR Conference, Baden-Baden, 1977, also Delft Hydraulics Laboratory, Publ. 191

7   KREISS, H.O., and OLIGER, J.,
    Methods for the approximate solution of time-dependent problems,
    World Meteor. Org., GARP Publ. Series no. 10, 1973

8   LEENDERTSE, J.J.,
    Aspects of a computational model for long-period water wave propagation,
    RAND Memorandum RM-5294-PR, 1967

9   MAHMOOD, K., and YEVJEVICH V. (eds.),
    Unsteady flow in open channels,
    Water Res. Publ., Fort Collins, Col., 1975

10  PAROT, J.M.,

Quelques aspects des écoulements bicouches quasi-horizontaux et de leur calcul,

La Houille Blanche 31, 1 (1976), 53-58


11  PERRELS, P.A.J., and KARELSE, M.,

A two-dimensional numerical model for salt intrusion in estuaries,

in: J.C.J. Nihoul (ed.) - Hydrodynamics of estuaries and fjords, Elsevier (1978),

also Delft Hydraulics Laboratory, Publ. 177 (1977)


12  SUNDSTRÖM, A.,

Boundary conditions for limited area integration of the viscous forecast equations,

Beiträge zur Physik der Atmosphäre, 50 (1977), 218-224


13  TRACOR, Inc.,

Estuarine modelling: an assessment,

Water Quality Office, Environmental Protection Agency, 1971


14  VREUGDENHIL, C.B.,

Two-layer shallow-water flow in two dimensions, a numerical study,

Submitted for publication in J. Comp. Phys., 1978


15  VREUGDENHIL, C.B., and VOOGT, J.,

Hydrodynamic transport phenomena in estuaries and coastal waters, scope of mathema-

tical models,

ASCE Symp. Modelling '75, San Francisco, 1975, also Delft Hydraulics Laboratory,

Publ. 155

FINITE ELEMENTS, A FLEXIBLE TOOL FOR
MODELLING ESTUARINE PROCESSES

K.-P. Holz

Chair of Fluid Mechanics
Technical University Hannover
Federal Republic of Germany

## Summary

During recent years, more and more finite element applications have
been made to simulate tidal wave propagation in oceans and estuaries.
Various types of formulations have been developed, aiming at an improvement in accuracy and economics of this technique. A review of recent tendencies will be given, and some numerical examples will show
the effectiveness of finite element approaches.

## Introduction

Finite element formulations have been continuously increasing in importance for the modelling of tidal process. Since the first applications of this technique were made a few years ago [1,2,3], rapid advances have been made. The first models started from implicit formulations, basing on the Galerkin approach. Meanwhile great attempts have
been made to reduce their high computational costs. This led to hybrid
and to explicit formulations which can be set up in a great variety
[4,5,6]. A review of this development will be given, some general aspects be pointed out, and some numerical applications shown.

## Formulation of the Problem

The describing equations for tidal processes can be derived directly
from the conservation equations of physics. For a finite volume element, the conservation of momentum yields

$$\frac{D}{Dt} \int_V v_i \, \rho \, dv - \int_V f_i \, \rho \, dv - \int_A \tau_{ij} \, u_j \, dA = 0 \qquad (1)$$

and the conservation of mass

$$\frac{D}{Dt} \int_V \rho \, dv = 0 \qquad (2)$$

51

The notation is as follows: i = 1,2,3

$\rho$ = density $\quad\quad\quad\quad\quad\quad\quad$ $v_i$ = components of velocity

$f_i$ = mass forces $\quad\quad\quad\quad\quad$ V = volume

$\tau_{ij}$ = surface stresses $\quad\quad\quad$ A = surface

With the commonly made assumptions of hydrostatic pressure distribution over the depth, neglection of surface pressure and of tide generating forces, and after introducing the Boussinesq approximation for the description of turbulent motion, the equations for shallow water-waves are obtained in an integral form.

$$\int \{ q_{\alpha,t} + (q_\alpha v_\beta)_\beta + \Omega \varepsilon_{\alpha\beta} q_\beta + g(a+h)\, h_{,\alpha} + \frac{\lambda}{a+h} \sqrt{v_{\beta,\beta}}\, q_\alpha$$

$$- \frac{\mu}{(a+h)} \sqrt{w_{\beta\beta}}\, (a+h) w_\alpha - A_H q_{\alpha,\beta\beta} \} \; dA = 0 \tag{3}$$

$$\int \{ h_{,t} + q_{\alpha,\alpha} \}\; dA = 0 \tag{4}$$

The quantity $q_\alpha$ stands for the flux, $w_\alpha$ for the wind components, $\mu$ and $\lambda$ are parameters for the wind and bottom friction terms. Moreover, $\Omega$ is the Coriolis parameter and $\varepsilon_{\alpha\beta}$ stands for

$$\varepsilon_{\alpha\beta} \quad \begin{vmatrix} 0 & -1 \\ 1 & 0 \end{vmatrix}, \quad \begin{matrix} \alpha = 1,2 \\ \beta = 1,2 \end{matrix} \tag{5}$$

$A_H$ is the eddy viscosity and A the area in the horizontal plane. The coordinate system is given in Fig. 1.

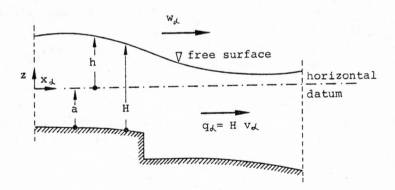

Fig. 1.: Notation

For decreasing volume, the formulation (3,4) leads to the differential equations for shallow-water waves which just are the kernel of the integrals.

Both formulations, the differential and the integral one, can now be used as basic equations for numerical solution procedures (Fig. 2).

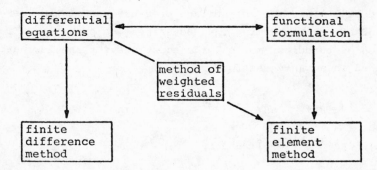

Fig. 2.: Solution Techniques

The finite difference method starts from a direct approach to the set of differential equations. If the finite element method, which basically is an integral method, is to be applied, the differential equations have to be put into an integral form. In most cases this is done by using the method of weighted residuals, which serves to minimize the error between the exact and the approximate solutions in an integral sense. A variety of methods can be applied, out of which the Galerkin approach is commonly chosen.

There still remains, however, some uncertainty as to whether this or another approach would be the best one. This problem can be overcome by setting up a variational functional for the set of differential equations. Attempts have been made but turned out not very successful [7,8]. But it even seems not to be necessary to go this way and make the problem more complicated. Equations (3,4) were obtained in an integral form and can be integrated directly. This procedure can be interpreted as a method of weighted residuals and corresponds to a subdomain approach with a constant weighting factor w. To show this, equations (3,4) are re-written in matrix form

$$\int (\underline{D}\ \underline{z} - \underline{r})\ dA = 0 \qquad\qquad (6)$$

$\underline{D}$ contains the differential operators, $\underline{z}$ the quantities h and $q_\alpha$, and $\underline{r}$ stands for the inhomogeneous part of the equations. If now an approximate solution $\underline{z}^*$ is set up, the error between this and the exact solution should be zero. This equally holds for the space and for the time domian

$$\iint (\underline{D}\ \underline{z} - \underline{r})\ dA\,dt - \iint w(\underline{D}\ \underline{z}^* - \underline{r})\ dA\,dt = 0 \tag{7}$$

From this the subdomain formulation follows directy:

$$\iint w(\underline{D}\ \underline{z}^* - \underline{r})\ dA\,dt = 0 \tag{8}$$

It ensures mass and momentum conservation in an integral sense and thus fulfils one of the basic requirements which have to be made for any numerical scheme.

Some remarks have still to be made concerning the boundary conditions. Equations (3,4) hold for a finite volume element. If the flux across its boundary is known in terms of the mass flux $\bar{q}$ normal to the boundary, and in terms of the momentum flux components $\bar{m}_\alpha$, they can be prescribed by using boundary integrals. For the mass flux the condition

$$\oint (q_\alpha u_\alpha - \bar{q})\,ds = 0 \tag{9}$$

and for the momentum flux the condition

$$\oint (v_\beta q_\alpha - \bar{m}_\alpha)\ n_\beta\ ds = 0 \tag{1o}$$

are obtained. $n_\alpha$ stands for the direction cosinus. As the differential equations are of second order, an additional condition for the turbulent exchange on the boundary has to be defined. This is done by the equation

$$\oint (A_H q_{\alpha,\beta} - \bar{t}_\alpha)\ _\beta\ ds = 0 \tag{11}$$

in which $\bar{t}_\alpha$ stands for the friction on the coastline which must be prescribed. Parametric formulations, set up in analogy to the bottom friction, have been successfully used [9].

The conditions (9,1o,11) allow for specifying constant inflow. So the discharge from rivers can easily be taken into account. Prescribed zero inflow defines a slip condition along the coastline. On the open sea side, however, normally neither the momentum flux nor the discharge are known. The only quantity available there is the water level h. This condition has to be directly implemented into the solution as an artificial boundary condition.

## Finite Element Discretization

In the finite element technique the system which has to be analyzed is discretized into arbitrary finite elements, which generally are triangular. Some authors [ 2 ] applied isoparametric elements as well. All these element shapes allow for a high adaptation to irregular boundaries and complex bathymetry. This is demonstrated by figures 3 and 4 which show an area in the German bight. The discretization fits extremely well to the deep shipping channel and to the intertidal flats.

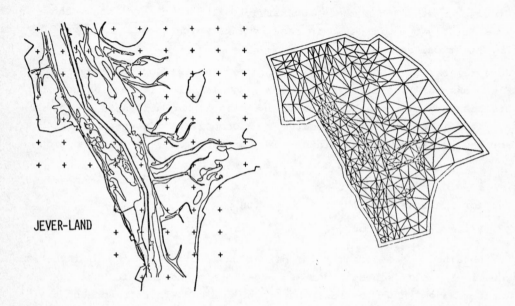

JEVER-LAND

Fig. 3: Topography of  Fig. 4: Discretization for
        Jade Estuary              Jade Estuary

## Implicit Models

The first step when setting up a numerical model by the finite element method is to choose a set of trial functions for the unknown water-level h and fluxes $q_\alpha$ on an element level.

$$q_\alpha = \phi_i \ (x_1, \ x_2) \ \hat{q}_{\alpha i} \tag{12}$$

$$h = \psi_i \ (x_1, \ x_2) \ \hat{n}_i \tag{13}$$

$\phi$ and $\psi$ describe the continuous distribution of h and q over the inner part of an element. The shape of this distribution depends on the values of the nodal parameters $\hat{q}_{\alpha i}$ and $\hat{n}_i$. As they are also belonging to neighbouring elements, they match the distribution from one element to the next and thus over the entire domain of solution.

The functions $\phi_i$ and $\psi_i$ have to be chosen in such a way that compatibility for the water-level and the fluxes is ensured also across the element boundaries. In most cases linear functions are used when the second order term in equation (3) is either omitted or removed by partial integration. Otherwise, higher order functions become necessary  2, 1o .

The integration of equation (8) is then performed on an element level, giving the matrix equation

$$\int (\underline{n} \left| \frac{q_\alpha}{h} \right|_t + \underline{m} \left| \frac{q_\alpha}{h} \right| - \underline{r}) \; dt = 0 \tag{14}$$

It is assumed here that the differential equations have been linearized. The discretization with respect to time can be performed in the same manner, or else by simply applying the Crank-Nicolson formula with the factor $\theta$ weighting the time levels. This then leads to

$$(\underline{n} + \Delta t\theta \quad \underline{m}) \left| \frac{q_\alpha}{h} \right|_{t+\Delta t} = (\underline{n} - \Delta t(1-\theta)\underline{m}) \left| \frac{q_\alpha}{h} \right|_t + \Delta t \; \underline{r} \tag{15}$$

These equations, which are set up for all elements, are then assembled to give a system of equations for the unknown nodal quantities.

$$(\underline{N} + \Delta t\theta\underline{M}) \left| \frac{Q_\alpha}{H} \right|_{t+\Delta t} = (\underline{N} - \Delta t(1-\theta)\underline{M}) \left| \frac{Q_\alpha}{H} \right|_t + \Delta t \; \underline{R} \tag{16}$$

It is non-symmetric, is structured as band matrix, and can be solved by the Gaussian algorithm.

It is quite obvious that this finite element approach leads to large equation systems. This disadvantage can be reduced by eliminating the flux $q_\alpha$ on an element level. Then out of the mixed model (16) a hybrid model [9,11] is generated, in which the nodal water-levels remain unknown

$$\underline{K} \; \underline{H}_{t+\Delta t} = \underline{F} \tag{17}$$

Now only one third of unknown parameters is contained in the equation system and the formulation has become more economic and attractive.

Stability problems do not occur in both formulations as long as the Crank-Nicolson factor is chosen    $\theta \geq 1/2$. This has been shown for the mixed model by numerical tests, and for the hybrid formulation by theoretical considerations [12].

## Explicit Models

In many situations, as for example the calculation of systems with intertidal flats, it becomes necessary to choose rather small time-steps
for the computation, which makes implicit models extraordinarily expensive. Moreover, for implicit models, a local refinement of the discretization makes a re-numbering of all models necessary when the band-
width of the equation system has to be minimized. Both arguments lead
to the development of explicit formulations.

One type of explicit models applies the lumping technique. Equation (15)
leads, with $\Theta = o$, to the formulation

$$\underline{N}\left|\begin{matrix} Q_\alpha \\ \overline{H} \end{matrix}\right|_{t+\Delta t} = (\underline{N} - \Delta t\ \underline{M})\ \left|\begin{matrix} Q_\alpha \\ \overline{H} \end{matrix}\right|_t + \Delta t\ \underline{R} \tag{18}$$

The matrix $\underline{N}$ follows from the integration of the terms $q_{\alpha,t}$ and $h_{,t}$ of
the differential equations. Each line of $\underline{N}$ describes the momentum and
mass distribution in the elements around their common central node.
If now mass and momentum of these elements are assumed to be concentrated in the central node itself, the matrix $\underline{N}$ becomes a diagonal
form $N_L$, and for each node an independent equation, and thus an explicit formulation, is obtained. This lumping technique was first
mentioned in [4]. It is of first order accuracy in the form of equation (18). Higher order schemes have been set up in analogy to the
leap-frog formulation [5] and in analogy to the LAX-WENDROFF-scheme [6].

Another approach to obtain explicit models was given in [13]. Here triangular finite elements in space and time were applied for the calculation of open channel flow. An extension of this approach to two-
dimensional situations leads to the use of a tetrahedron in space

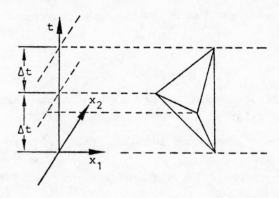

Fig. 5: Explicit Tetrahedron Element in Space and Time

and time (Fig. 5). These elements are of second order accuracy. They were applied to the system (Fig. 4). Results are given in the next chapter.

The third approach for generating explicit elements starts from the hybrid implicit formulation (17), in combination with a subdomain approach [11]. The weighting functions are retained in analytical form in the equation system, and then determined in a way that the equation system decouples into diagonal form. It becomes obvious that as a condition for this, the weighting should be extended only over the domain of influence of the characteristics around a nodal point (Fig. 6).

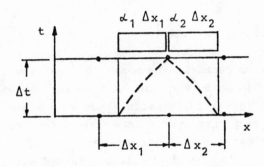

Fig. 6: Weighting for Explicit Formulations

All explicit models depend for stability on the COURANT-number

$$\Delta t \leq \frac{\Delta x}{\sqrt{2g(a+h)}} \tag{19}$$

Though this enforces a severe restriciton on the time-step, there are many situations where they are more economic than implicit models.

Numerical Results

When developing a new numerical method, quite a number of basic tests have to be run before a scheme is ready for application. A few results from these tests are given here. The first test concerns a comparison between the Galerkin approach and the direct integration of equ. (3,4). The example is taken from an open channel flow calculation with a mixed implicit formulation. The system and the results are shown in Fig. 7.

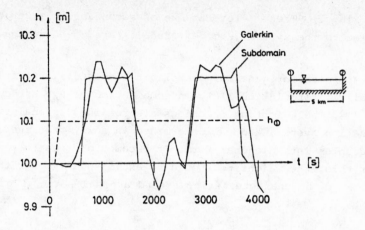

Fig. 7: One-dimensional Calculation

In a straight rectangular channel, the water-level at its left end is raised from 1o.oo m to 1o.1o m. The time history for the water-level variation is plotted (Fig. 7) for a calculation with 5 elements of 1 km length. The time-step corresponds to a Cournat-number of one. The Galerkin formulation shows a rather noisy behaviour and no exact reproduction of the phase, whereas the results obtained with the direct integration are much more satisfactory.

The second test was run in order to find out whether a rotational motion is reproduced correctly. The calculation is performed for a schematic estuary, for which experimental data were available too. The system and its discretization is shown in figure 8. The results, given by figure 9, are obtained by the explicit model using tetrahedral elements in space and time. The eddy which is generated behind the re-entrant corner during flood-phase is fairly well reproduced, and in good agreement with the experiment. For the same discretization a comparison with respect to computing time was made between a mixed implicit and the explicit model. It was shown that, on a time-step basis, the explicit formulation was faster by a factor of 3o, but still slower by a factor of 2 in comparison to a finite difference formulation using a staggered net. A further test between implicit mixed models, one using the Galerkin technique, the other the direct integration, showed practically no influence of the weighting technique on the calculated water-levels, the calculated velocities, however, differed considerably in some areas of the system.

The last example is taken from an application to the system given in Figure 4. The explicit tetrahedral model was used . Figure 1o shows a typical velocity distribution.

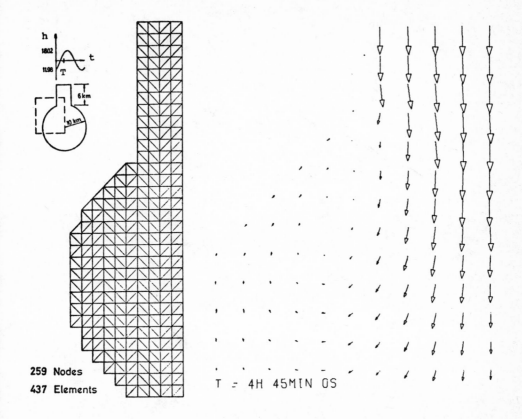

259 Nodes
437 Elements

T = 4H 45MIN 0S

Fig. 8: Discretization     Fig. 9: Calculated Velocities at time t
for Schematic
Estuary

## Conclusions

A review of finite element models for the calculation of shallow water
waves is given. It is shown that the method of weighted residuals,
when using a subdomain approach, satisfies conservation of mass and
momentum in an integral sense. Various implicit and explicit formulations are presented. A few numerical examples demonstrate that good
agreement with observations can be obtained by finite element models.
By comparing the computation time of an explicit finite element model

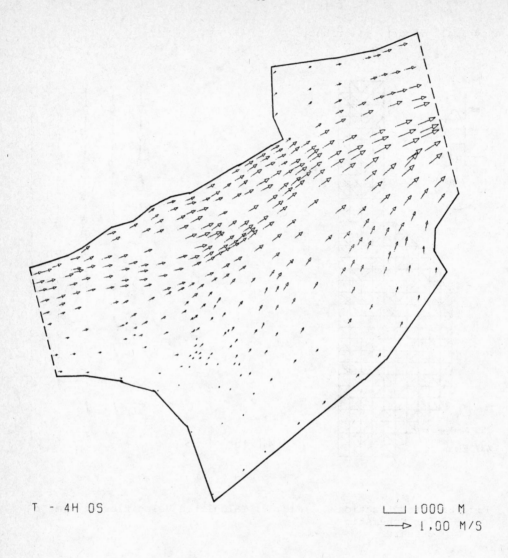

T - 4H 0S    └─┘ 1000 M
             ──▷ 1.00 M/S

Fig. 1o: Velocities after slack water

with a corresponding finite difference model, it is found that the
latter is still faster, whereas the finite element formulation offers a
much higher flexibility. This has naturally to be paid for. Generally
speaking, the finite element method has become as valuable a technique
for modelling estuarine processes as the finite difference method is
nowadays.

References

[1]   GROTKOP, G., Finite element analysis of long period water waves,
      Comp. Meths. Appl. Mech. Eng. 2 (1973), pp 147 - 157

[2]   TAYLOR, C., DAVIS, J.M., Tidal propagation and dispersion in
      estuaries, in: GALLAGHER, R.H., ODEN, J.T., TAYLOR, C.,
      ZIENKIEWICZ, O.C., (eds.), Finite Elements in Fluids,
      Wiley, London 1975, pp. 95 - 118

[3]   CONNOR, J.J., WANG, J., Finite element modelling of hydrodynamic
      circulation, in: BREBBIA, C.A., CONNOR, J.J., Numerical
      Methods in Fluid Dynamics, Pentech Press, London, 1974,
      pp. 355 - 387

[4]   WANG, H.P., Multi-levelled finite element hydrodynamic model of
      Block Island Sound, in: GRAY,W.G., PINDER, G.F.,
      BREBBIA, C.A., Finite Elements in Water Resources, Pentech
      Press, London, 1977, pp. 4.69 - 4.93

[5]   KAWAHARA, M., TAKEUCHI, N., YOSHIDA, T., Two step explicit ele-
      ment method for Tsunami wave propagation analysis, Int.
      J. Num. Meths. Eng. 12, 1978, pp. 331 - 351

[6]   HOLZ, K.-P., Explizite Finite Element Formulierung zur Berech-
      nung langperiodischer Flachwasserwellen, ZAMM 58,
      pp. 227 - 278

[7]   GUYMON, G.L., Finite element solution for general fluid flow,
      J. Hydr. Div. 6, 1973, pp. 913 - 919

[8]   HOLZ, K.-P., Numerische Simulation von Flachwasserwellen mit der
      Methode der finiten Elemente, Fortschritt-Berichte,
      Reihe 4, Nr. 28, Verein Deutscher Ingenieure (VDI),
      Düsseldorf, 1976

[9]   HERRLING, B., Eine hybride Formulierung in Wasserständen zur
      Berechnung von Flachwasserwellen mit der Methode der
      finiten Elemente, Fortschritt-Berichte, Reihe 4, Nr. 37,
      Verein Deutscher Ingenieure (VDI), Düsseldorf, 1977

[1o]  GRAY, W.G., An efficient finite element scheme for two-dimensio-
      nal surface water computation, in: GRAY, W.G., PINDER, G.F.,
      BREBBIA, C.A., Finite Elements in Water Resources, Pentech
      Press, London, 1977, pp. 4.33 - 4.49

[11]  MEISSNER, U., An explicit-implicit water-level model for tidal
      computations of rivers, Comp. Meths. Appl. Mech. Eng. 13
      (1978), pp. 221 - 232

[12]  MEISSNER, U., Discretization techniques and time integration
      schemes for hydrodynamical finite element models, in:
      BATHE, K.-J., ODEN, J.T., WUNDERLICH, W., Formulations
      and Computational Algorithms in Finite Element Analysis,
      MIT-Press, Boston, 1976, pp. 1o12 - 1o38

[13]  HOLZ, K.-P., HENNLICH, H.H., Numerical experience from the com-
      putation of tidal waves by the finite element method,
      in: GRAY, W.G., PINDER, G.F., BREBBIA, C.A., Finite
      Elements in Water Resources, Pentech Press, London, 1977,
      pp. 4.19 - 4.31

# MATHEMATICAL MODELLING OF FLUID FLOW
## USING THE BOUNDARY ELEMENT METHOD

C.A. Brebbia and L.C. Wrobel
Department of Civil Engineering
Southampton University, U.K.

## 1.  INTRODUCTION

Boundary solutions in the past have been used almost exclusively in solid mechanics [1]. However, they can be of considerable interest for many fluid problems [2]. This paper attempts to review the fundamentals of a boundary solution technique known as the boundary element method and indicates how it can be applied in practical cases.

In the boundary element method the external surface of the domain is divided into a series of elements over which the functions under consideration are assumed to vary in much the same way as in finite elements. This produces a series of nodal unknowns on the surface of the body only. These unknowns are related through the same influence functions used in the boundary integral equation method. The capability of using different types of surface elements is important as in the past, integral equation formulations were generally restricted to constant sources assumed to be concentrated at a series of points on the boundary. Furthermore, the technique is now being re-examined using weighted residual type statements, which allows for the technique to be more easily extended to non-linear, time dependent and other complex problems.

This paper starts explaining how boundary elements can be applied to solve time independent potential problems. The methodology is presented in matrix form for the Poisson's equation considering first steady state boundary conditions including free surface and afterwards extending them to cover time dependent conditions such as those occurring during drawdown. Time dependent problems are more fully discussed in section 3, where a time dependent potential problem is presented. For this case the starting weighted residual statement has to include the initial in addition to the boundary conditions. This type of relation implies a new type of fundamental solution, which depends on time as well as the spatial coordinates. The approach indicated in this paper is of considerable interest as it eliminates the need of integrating step by step in time using finite difference or similar discretizations.

## 2.  POTENTIAL PROBLEMS

Let us first consider a potential function u over a domain $\Omega$, where the following governing equation is to be satisfied,

$$\nabla^2 u - p = 0 \qquad \text{in } \Omega \tag{1}$$

The boundary conditions corresponding to this problem are of two types: *essential* conditions, such as $u = \bar{u}$ on $\Gamma_1$ and *natural* conditions such as $q = \frac{\partial u}{\partial n} = \bar{q}$ on $\Gamma_2$. The total boundary is $\Gamma = \Gamma_1 + \Gamma_2$.

For our numerical solution u will be approximated and we can minimize the error thus introduced by weighting the governing equation by a new function u*. This gives,

$$\int_\Omega (\nabla^2 u - p) u^* \, d\Omega = \int_{\Gamma_2} (q - \bar{q}) u^* \, d\Gamma - \int_{\Gamma_1} (u - \bar{u}) q^* \, d\Gamma \qquad (2)$$

where $q^* = \frac{\partial u^*}{\partial n}$ .

After integrating by parts twice the terms in the Laplacian equation (2) becomes,

$$\int_\Omega (\nabla^2 u^*) u \, d\Omega - \int_\Omega p u^* \, d\Omega = - \int_{\Gamma_2} \bar{q} u^* \, d\Gamma - \int_{\Gamma_1} q u^* \, d\Gamma + \int_{\Gamma_2} u q^* \, d\Gamma + \int_{\Gamma_1} \bar{u} q^* \, d\Gamma \quad (3)$$

The function u* is now assumed to be the fundamental solution of the equation, representing a concentrated potential at a point i, i.e. the solution of

$$\nabla^2 u^* + \Delta_i = 0 \qquad (4)$$

where $\Delta_i$ is the Dirac delta function. Hence equation (3) can be written as,

$$u_i + \int_{\Gamma_2} u q^* \, d\Gamma + \int_{\Gamma_1} \bar{u} q^* \, d\Gamma + \int_\Omega p u^* \, d\Omega = \int_{\Gamma_2} \bar{q} u^* \, d\Gamma + \int_{\Gamma_1} q u^* \, d\Gamma \qquad (5)$$

It can be noted that for an isotropic three dimensional medium the fundamental solution of equation (4) is

$$u^* = \frac{1}{4\pi r} \qquad (6)$$

where r is the distance from the point of application of the concentrated potential to the point under consideration. For two dimensions the solution is,

$$u^* = \frac{1}{2\pi} \ln \left(\frac{1}{r}\right) \qquad (7)$$

Other fundamental solutions for potential problems are given in reference [3].

Equation (5) is valid for any point inside the domain, but in order to formulate the problem as a boundary technique one needs to take this point to the boundary. For this case equation (5) can be written as,

$$c_i u_i + \int_\Gamma u q^* \, d\Gamma + \int_\Omega p u^* \, d\Omega = \int_\Gamma q u^* \, d\Gamma \qquad (8)$$

Note that we have written the integrals for the whole $\Gamma = \Gamma_1 + \Gamma_2$ boundary but that, depending on which part of the boundary we consider, the $\bar{u}$ or $\bar{q}$ values will be known. We can introduce these boundary conditions at a later stage.

The value of $c_i$ is:

$c_i = 1$ for an internal point.

$c_i = 0$ for an external point.

$c_i = \frac{1}{2}$ for a boundary point on a smooth boundary.

Boundary Elements.

Equation (8) can now be applied on the boundary of the domain under consideration. This boundary can be divided into n elements. The points where the unknown values are considered are called 'nodes' and are similar to those of finite elements. The main difference is that now elements and nodes are defined only on the $\Gamma$ boundary. The functions u and q over each boundary element are given by,

$$u = \underline{\Phi}^T \underline{u}^n$$
$$q = \underline{\psi}^T \underline{q}^n$$

$$(9)$$

and equation (8) is discretized as follows,

$$c_i u_i + \sum_{j=1}^{n} \int_{\Gamma_j} uq^* \, d\Gamma + \sum_{k=1}^{m} \int_{\Omega_k} pu^* \, d\Omega = \sum_{j=1}^{n} \int_{\Gamma_j} qu^* \, d\Gamma \qquad (10)$$

Note that m internal elements or cells need to be defined to compute the integrals in $\Omega$ but these elements do not introduce any further unknown and hence the problem is still a boundary problem.

We can substitute the u and q values given by (9) into (10) and carry out the integrations (usually numerically). This gives for each node, after assembling, the following equation,

$$c_i u_i + \sum_{j=1}^{n} \hat{H}_{ij} u_j + B_i = \sum_{j=1}^{n} G_{ij} q_j \qquad (11)$$

or

$$\sum_{j=1}^{n} H_{ij} u_j + B_i = \sum_{j=1}^{n} G_{ij} q_j \qquad (12)$$

where the $B_i$ term is the result of having integrated the domain term and

$H_{ij} = \hat{H}_{ij}$ for $i \neq j$ and $H_{ij} = \hat{H}_{ij} + c_i$ for $i = j$.

The whole set can be written in matrix form as follows,

$$\underset{\sim}{H} \, \underset{\sim}{U} + \underset{\sim}{B} = \underset{\sim}{G} \, \underset{\sim}{Q} \qquad (13)$$

Note that $n_1$ values of u and $n_2$ values of q are known on the boundary, hence equation (13) can be reordered in such a way that all the unknowns are on the left hand side, i.e.

$$A \underset{\sim}{X} = \underset{\sim}{F} \qquad (14)$$

where X is the vector of unknowns u's and q's.

When the surface is not smooth at the point 'i' the $c_i = \frac{1}{2}$ value is no longer valid. However, we can always calculate the diagonal terms of $\underset{\sim}{H}$ by the fact that when a uniform potential is applied over the whole boundary, the normal derivatives must be zero. Hence equation (13) in the absence of body forces becomes,

$$\underset{\sim}{H} \underset{\sim}{U} = \underset{\sim}{0} \qquad (15)$$

where U is a uniform potential. Thus the sum of all the elements of $\underset{\sim}{H}$ in any row ought to be zero and the value of the coefficient on the diagonal can be easily calculated once the off-diagonal coefficients are all known, i.e.

$$H_{ii} = - \sum_{\substack{j=1 \\ (i \neq j)}}^{n} H_{ij} \qquad (16)$$

Once the values of the boundary unknowns are found, the values of u and q anywhere can be calculated using equation (8). The internal fluxes are obtained by computing the derivatives of (8).

Free Surface Condition.

Boundary conditions such as free surfaces can, in many cases, be treated simply by approximating the location of the surface, which is not known a priori, and iterating until the free surface condition is fulfilled.

In cases of seepage the condition is that at any point on the free surface the potential head u equals the elevation head from a fixed reference plane. This condition is also valid for cases such as flow over a spillway when the velocity head can be neglected, i.e., when the height of water over the nappe is small by comparison with the spillway height [4].

The problem is governed by a Laplace's equation for the potential u. For the solution, an initial guess is assumed for the free surface and the problem is solved for u applying the condition q = 0 (there is no flux through the free surface). The calculated potential at every nodal point on the free surface is then compared against its elevation; if the difference between these two values is greater than a maximum acceptable error, this difference is algebraically added on the elevation of the nodal point, and a new iteration is carried out.

The elements on the G and H matrices – equation (13) – corresponding to the influence of fixed boundary nodes on other fixed boundary nodes will remain constant during the analysis, hence they can be computed once and stored. The potentials at internal points are calculated just after the correct position of the free surface has been found. With finite elements instead the internal potentials need to be computed during all iterations.

For unconfined transient groundwater flow the method can also be used. For this case the kinematic condition on the free surface is [5],

$$q_2 = \frac{\partial \eta}{\partial t} + q_1 \frac{\partial \eta}{\partial x_1} \tag{18}$$

where $q_1$, $q_2$ are the velocities in the $x_1$, $x_2$ directions (figure 1) and $\eta$ is the elevation of the free surface with relation to an arbitrary plane. We can write,

$$\frac{\partial \eta}{\partial t} = q_2 - q_1 \frac{\partial \eta}{\partial x_1} \tag{19}$$

From geometric considerations we have that,

$$\tan \beta = - \frac{\partial \eta}{\partial x_1} \tag{20}$$

in which $\beta$ is the angle the free surface makes with the $x_1$ axis. Hence one obtains

$$\frac{\partial \eta}{\partial t} = - \frac{q_n}{\cos \beta} \tag{21}$$

where $q_n = \frac{\partial u}{\partial n}$ is the normal velocity.

At the free surface we still have the condition $u = \eta$. Equation (21) becomes,

$$\frac{\partial u}{\partial t} = - \frac{q_n}{\cos \beta} \tag{22}$$

This equation can be written in finite difference form as,

$$u^{k+1} = u^k - \frac{\Delta t}{\cos \beta^k} \left[ \theta . \, q_n^{k+1} + (1 - \theta) q_n^k \right] \tag{23}$$

$\theta$ is a weighting factor that positions the derivative between the time levels $k \, \Delta t$ and $(k+1) \, \Delta t$. In the equation, the angle $\beta$ is computed at level $k$ even though the equation is written for the level $k+1$. Although this problem can be avoided by iteration, a small time step provides sufficient accuracy [6].

As an example to show the way the boundary condition given by equation (23) is introduced into the analysis, suppose the problem under consideration has solid boundaries (where $q_n = 0$) and one free surface. We assume that the solution is known at time $k \, \Delta t$. For a Laplace's equation the boundary technique gives, after assembling, a system of equations as follows,

$$\underset{\sim}{H} \, \underset{\sim}{U} = \underset{\sim}{G} \, \underset{\sim}{Q} \tag{24}$$

This system is rearranged in order that all the unknowns are on the left hand side,

$$\begin{bmatrix} \underset{\sim}{H}_S & -\underset{\sim}{G}_F \end{bmatrix} \begin{Bmatrix} u_S \\ q_F \end{Bmatrix}^{k+1} = \begin{bmatrix} \underset{\sim}{G}_S & -\underset{\sim}{H}_F \end{bmatrix} \begin{Bmatrix} q_S \\ u_F \end{Bmatrix}^{k+1} \tag{25}$$

where the subscript S means the solid boundaries and F the free surface. Substituting $u_F^{k+1}$ by its value on equation (23) one obtains,

$$\begin{bmatrix} \underset{\sim}{H}_S & \underset{\sim}{H}_F \dfrac{\Delta t . \theta}{\cos\,\beta} - \underset{\sim}{G}_F \end{bmatrix} \begin{Bmatrix} u_S \\ q_F \end{Bmatrix}^{k+1} = \begin{bmatrix} \underset{\sim}{G}_S & -\underset{\sim}{H}_F \end{bmatrix} \begin{Bmatrix} q_S \\ u_F^k - \dfrac{\Delta t}{\cos\,\beta}(1-\theta)q_F^k \end{Bmatrix} \tag{26}$$

After solving equation (26) for $q^{k+1}$, we can use equation (23) to find $u^{k+1}$ and the computation cycle is completed, so the solution can be advanced in time.

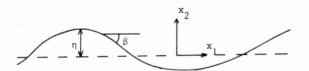

Figure 1. Geometrical definitions for free surface

## 3. TIME DEPENDENT PROBLEMS

In section 2 we have already seen how problems with time dependent boundary conditions can be treated. Here we will consider the case of having a time dependent potential problem governed by the following equation,

$$\nabla^2 u = \frac{\partial u}{\partial t} \tag{27}$$

with boundary conditions of the same type as previously seen and some initial conditions. As the problem is now time dependent, the equation will also be integrated with relation to time. Weighting expression (27) we have,

$$\int_o^t \int_\Omega \left(\nabla^2 u - \frac{\partial u}{\partial t}\right) u^* \, d\Omega \, d\tau = \int_o^t \int_{\Gamma_2} (q - \bar{q})u^* \, d\Gamma \, d\tau - \int_o^t \int_{\Gamma_1} (u - \bar{u})q^* \, d\Gamma \, d\tau \tag{28}$$

where $0 \le \tau \le t$.

Integrating by parts twice we find,

$$\int_o^t \int_\Omega \left(\nabla^2 u^* + \frac{\partial u^*}{\partial t}\right) u \, d\Omega \, d\tau - \left[\int_\Omega u \, u^* \, d\Omega\right]_{\tau=o}^{\tau=t} + \int_o^t \int_\Gamma q u^* \, d\Gamma \, d\tau = \int_o^t \int_\Gamma u q^* \, d\Gamma \, d\tau \tag{29}$$

where the $\frac{\partial u^*}{\partial t}$ term was obtained integrating by parts with respect to time.

The corresponding fundamental solution for this equation is [7],

$$u^* = \frac{1}{(4\pi\xi)^{d/2}} \exp\left(\frac{-r^2}{4\xi}\right) \tag{30}$$

where $\xi = t-\tau$ and d is the number of spatial dimensionality, i.e., d = 3 for three dimensional problems, etc. The fundamental solution possesses the properties,

$$\nabla^2 u^* + \frac{\partial u^*}{\partial t} = 0 \quad \text{in } \Omega \text{ for all } \tau \tag{31}$$

and for $\tau = t$:

$$\int_\Omega u^* \, d\Omega = \begin{cases} 0 & \text{for } r \neq 0 \\ 1 & \text{for } r = 0 \end{cases} \tag{32}$$

Substituting this solution into equation (29), one obtains for a point 'i',

$$u_i + \int_o^t \int_\Gamma u q^* \, d\Gamma \, d\tau = \int_o^t \int_\Gamma q u^* \, d\Gamma \, d\tau + \left[\int_\Omega u u^* \, d\Omega\right]_{\tau=o} \tag{33}$$

The last term in the above formula corresponds to the initial conditions at $\tau = 0$. Since the fundamental solution itself is time dependent, one does not need to propose an iterative scheme to solve time dependent problems as it is usually done in finite elements or finite differences.

EXAMPLE 1.  Lake Circulation

Flow in lakes can be approximated to provide an initial estimate of the circulation, which can then be checked against the full shallow water equations. This flow is governed by the following linearised equations, obtained by neglecting the inertia terms in the momentum equations [5],

$$- f \, q_2 + \rho \, g \, H \, \frac{\partial \eta}{\partial x_1} + (\tau_1^s - \tau_1^b) = 0$$

$$f \, q_1 + \rho \, g \, H \, \frac{\partial \eta}{\partial x_2} + (\tau_2^s - \tau_2^b) = 0 \tag{a}$$

and the continuity formula,

$$\frac{\partial q_1}{\partial x_1} + \frac{\partial q_2}{\partial x_2} = 0 \tag{b}$$

where:

    $f$ = Coriolis parameter

$q_1, q_2$ = vertically integrated velocity components in $x_1$, $x_2$ directions

    $\rho$ = mass density

    $g$ = acceleration of gravity

    $H = h + \eta$ = total depth of water

    $h$ = depth with relation to the mean water level

    $\eta$ = elevation of the free surface

    $\tau^s$ = wind stresses

    $\tau^b$ = bottom friction stresses

If the $\eta$ values are much smaller than the $h$ we can write $H \simeq h$, hence

$$- f\, q_2 + \rho\, gh\, \frac{\partial \eta}{\partial x_1} + (\tau_1^s - \tau_1^b) = 0$$

$$\tag{c}$$

$$f\, q_1 + \rho\, gh\, \frac{\partial \eta}{\partial x_2} + (\tau_2^s - \tau_2^b) = 0$$

Assuming the $\tau^b$ terms to be linearly proportional to the mean momentum components,

$$\tau_1^b = \gamma q_1 \quad , \quad \tau_2^b = \gamma q_2 \tag{d}$$

we can cross-differentiate equations (c) and afterwards subtract both equations. Assuming that the derivatives of $h$ are negligible (i.e. the bottom slope is small) this gives, taking continuity into consideration, the following equation,

$$\frac{\partial \tau_1^s}{\partial x_2} - \frac{\partial \tau_2^s}{\partial x_1} = \gamma \left( \frac{\partial q_1}{\partial x_2} - \frac{\partial q_2}{\partial x_1} \right) \tag{e}$$

One can propose a stream function $\psi$ such as,

$$q_1 = \frac{\partial \psi}{\partial x_2} \quad , \quad q_2 = - \frac{\partial \psi}{\partial x_1} \tag{f}$$

and formula (d) becomes,

$$\nabla^2 \psi = \frac{1}{\gamma}\, w(x_1, x_2) \tag{g}$$

where

$$w(x_1, x_2) = \frac{\partial \tau_1^s}{\partial x_2} - \frac{\partial \tau_2^s}{\partial x_1} \tag{h}$$

Note that we have included the Coriolis parameter but assumed it constant for all the lake, i.e. the lake is small enough to allow the neglect of local variations in the Coriolis forces. If we take,

$$X_1 = \frac{x_1}{L} \quad , \quad X_2 = \frac{x_2}{L}$$

$$W(X_1, X_2) = \frac{w(x_1, x_2)}{T/L} \tag{i}$$

$$\psi = \frac{\psi}{(f\varepsilon/2H^2)^{\frac{1}{2}} L^2}$$

L being the lateral characteristic length of the lake, T the characteristic wind stress and $\varepsilon$ the eddy viscosity coefficient, equation (g) takes the non-dimensional form,

$$\nabla^2 \psi = \frac{1}{\delta} W(X_1, X_2) \tag{j}$$

where

$$\delta = \frac{\gamma L (f\varepsilon/2)^{\frac{1}{2}}}{TH} \tag{k}$$

We analysed, using the above formulation, the wind circulation in Lagoa dos Patos, Brazil (figure 2-a). As a first numerical example, we calculate the stream lines for the flow in and out of the lake without wind effects, taking $\psi = 0$ for the west shore and $\psi = 1$ for the east shore. Results are shown in figure 2-b. For this case, the governing equation becomes a Laplace equation.

If we consider the right hand side of equation (j) equal to 1, $X_1$ and $X_2$, this allows for a superposition of three different sets of results in order to obtain any solution of the type,

$$\nabla^2 \psi = A + BX_1 + CX_2 \tag{$\ell$}$$

where the right hand side represents a quadratic wind stress distribution. This term is included in the analysis by dividing the domain into cells and integrating numerically over all the cells. Results are shown for a linear wind stress distribution, $A = -1$, $B = 0$, $C = 0$ (figure 2-c) and a quadratic wind stress distribution, $A = 1$, $B = -3$, $C = 0$ (figure 2-d).

EXAMPLE 2.  Flow in a Curved Channel

The channel showed in figure 3-a was analysed experimentally at the Department of Civil Engineering, University of Southampton. The dimensions of the channel are shown in the figure. The depth varies slightly over the region, from 10.12 cm to 12.53 cm.

In order to perform the solution using a two dimensional boundary element program, an averaged depth was considered. The flow was assumed to be potential, i.e. incompressible and inviscid. Hence the problem can be represented by a Laplace's equation for the stream function $\psi$. Results for the longitudinal velocity distribution in some cross-sections are compared against experimental velocities in figure 3-b and table I, showing good agreement although the experimental velocities are depth averaged.

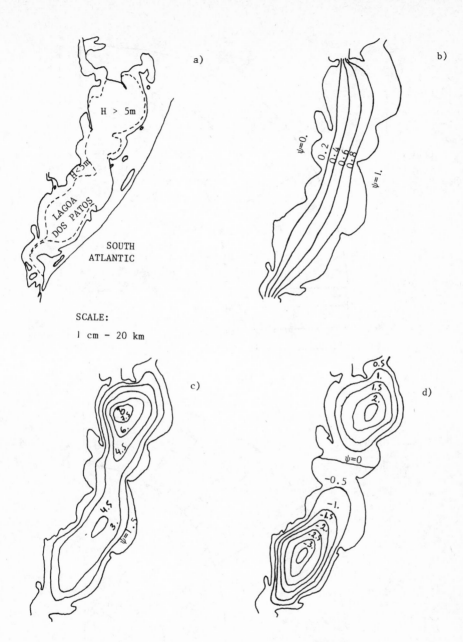

Figure 2. Lagoa dos Patos: a) Geometry; b) Flow pattern for potential
flow; c) Wind driven mean circulation pattern due to a linear
stress distribution; d) Wind driven mean circulation pattern
due to a quadratic stress distribution

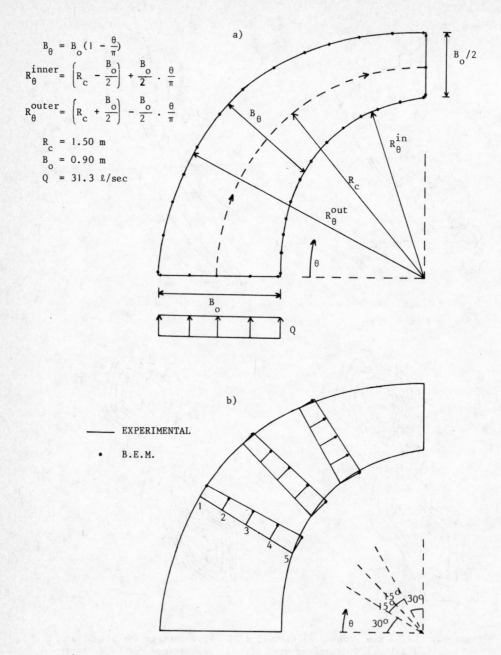

$$B_\theta = B_o \left(1 - \frac{\theta}{\pi}\right)$$

$$R_\theta^{inner} = \left(R_c - \frac{B_o}{2}\right) + \frac{B_o}{2} \cdot \frac{\theta}{\pi}$$

$$R_\theta^{outer} = \left(R_c + \frac{B_o}{2}\right) - \frac{B_o}{2} \cdot \frac{\theta}{\pi}$$

$$R_c = 1.50 \text{ m}$$
$$B_o = 0.90 \text{ m}$$
$$Q = 31.3 \text{ } \ell/\text{sec}$$

a)

b)

——— EXPERIMENTAL

• B.E.M.

Figure 3. Flow in a curved channel: a) Geometry and boundary elements discretization; b) Velocity distributions at some cross-sections

The method can be easily extended to study the three dimensional potential flow, in which case only surface elements are required.

| POINT | 30° | | 45° | | 60° | |
|---|---|---|---|---|---|---|
| | EXP. | B.E.M. | EXP. | B.E.M. | EXP. | B.E.M. |
| 1 | 0.29 | 0.28 | 0.35 | 0.31 | 0.43 | 0.36 |
| 2 | 0.33 | 0.31 | 0.38 | 0.35 | 0.47 | 0.41 |
| 3 | 0.37 | 0.35 | 0.42 | 0.39 | 0.49 | 0.44 |
| 4 | 0.41 | 0.39 | 0.46 | 0.43 | 0.52 | 0.47 |
| 5 | 0.46 | 0.45 | 0.51 | 0.50 | 0.57 | 0.54 |

Table I   Longitudinal velocities (m/sec) for curved channel

## 4. CONCLUSIONS

The main conclusion of this paper is that the boundary element method can be applied to solve many fluid mechanics problems and that it presents a series of advantages over more classical domain type techniques such as finite elements or finite differences.  One of the most interesting features of the technique is that a much smaller resulting system of equations and a considerable reduction in the data required to solve a problem can be achieved.  These advantages are more marked in three dimensional problems.  In addition, the numerical accuracy of boundary elements can be greater than that of techniques such as finite elements.

Boundary elements are also well suited for solving problems with infinite domains such as those frequently occurring in fluid mechanics for which the classical domain methods are obviously unsuitable.

## 5. ACKNOWLEDGEMENTS

The second author is indebted to C.N.Pq,Conselho Nacional de Desenvolvimento Cientifico e Tecnologico, Brasil, for the financial support provided during his studies at Southampton University.

## REFERENCES

1.  Brebbia, C.A. (editor),  "Recent Advances in Boundary Element Methods", Southampton, 1978, Pentech Press.
2.  Brebbia, C.A. and Wrobel, L.C.,  "Applications of Boundary Elements in Fluid Flow", Second Int. Conference on Finite Elements in Water Resources, Imperial College, London, 1978, Pentech Press.

3.  Brebbia, C.A., "The Boundary Element Method for Engineers", Pentech Press, London, 1978.

4.  Chow, V.T., "Open-Channel Hydraulics", McGraw-Hill, U.S.A., 1959.

5.  Connor, J.J. and Brebbia, C.A., "Finite Element Techniques for Fluid Flow", Newnes-Butterworths, 1976.

6.  Liu, P.L-F. and Liggett, J.A., "Boundary Integral Solutions to Groundwater Problems", First Int. Conf. on Applied Numerical Modelling, Southampton, 1977.

7.  Wu, J.C., "Finite Element Solution of Flow Problems using Integral Representations", Second Int. Symp. on Finite Element Methods in Flow Problems, Sta. Margherita, Italy, 1976.

Spectral method for the numerical solution of the three-dimensional
hydrodynamic equations for tides and surges

by

N. S. Heaps

Institute of Oceanographic Sciences, Bidston Observatory, England.

## Abstract

A review is given of the mathematical aspects of a spectral method developed for
the numerical solution of the three-dimensional hydrodynamic equations for tides
and surges.  Features of the method not given proper emphasis before are described
and the use of different frictional conditions at the sea bed is discussed.

## Introduction

During recent years a spectral method has been developed for the numerical
solution of the three-dimensional hydrodynamic equations for tides and surges
(Heaps 1972, 1973, 1974, 1976;  Heaps and Jones 1975, 1977;  Clarke 1974).  The
horizontal components of current are expanded in terms of eigenfunctions through
the depth.  Coefficients of these expansions varying in the horizontal and through
time, are determined from a two-dimensional numerical time-stepping procedure.  In
this way the three-dimensional current structure is computed as time advances, also
the changing pattern of sea-surface elevation.

In this paper the basic mathematical theory of the above method, as it presently
stands, is reviewed and some new comparisons are made in its development with
differing forms of frictional boundary condition at the sea bed.  Also more emphasis
is laid than hitherto on the completion of the expansions to infinity in order to
ensure the explicit satisfaction of surface stress and bottom stress boundary
conditions.

For simplicity only the linearized forms of the hydrodynamic equations are
considered.  So far most of the work has in fact been concerned with these.  The use
of the method with nonlinear equations has begun (Heaps 1976) but is in its early
stages.  All the theory is concerned with the motion of homogeneous water.

Further developments by Davies (1977a, 1977b, 1978) have used B-spline and
cosine-function representations of the vertical profile of current, solving the
three-dimensional hydrodynamic equations employing a Galerkin-type procedure.
Generally there has been good agreement between comparable results obtained from
this and the earlier work.  The Galerkin approach relies on an arbitrary choice of
base functions for the vertical expansion of current and, with a _finite_ expansion,
can satisfy the surface and bottom boundary conditions exactly.  On the other hand,
the eigenfunction approach itself determines the vertical expansion of current and
the completion of this to infinity ensures the exact satisfaction of the surface and
bottom conditions.  The Galerkin approach may include an expansion of eddy viscosity
through the depth and in this way can allow for fairly general variations in this

parameter. Most of the work with eigenfunctions has been carried out, as described in the present paper, with an eddy viscosity uniform through the depth but there are possibilities for a relaxation of this restriction to include an element of depth-dependency.

## Notation

Denote by

$t$    the time ;

$x, y, z$    Cartesian coordinates, forming a left-handed set, in which $x, y$ are measured in the horizontal plane of the undisturbed sea surface and $z$ is depth below that surface ;

$h$    the undisturbed depth of water ;

$\zeta$    the ratio $z/h$ ;

$\varsigma$    the elevation of the sea surface ;

$\bar{\varsigma}$    the equilibrium tide ;

$u, v$    the components of current at depth $z$ in the directions of increasing $x, y$ respectively ;

$\bar{u}, \bar{v}$    the depth-mean values of $u, v$ :

$$\bar{u} = \frac{1}{h} \int_0^h u\, dz \quad, \quad \bar{v} = \frac{1}{h} \int_0^h v\, dz \quad;$$

$F_0, G_0$    the components of wind stress at the sea surface in the $x, y$ directions ;

$F_h, G_h$    the components of frictional stress of the water on the sea bed in the $x, y$ directions ;

$p_a$    the atmospheric pressure on the water surface ;

$\rho$    the density of the water, assumed to be uniform and constant ;

$\gamma$    the geostrophic coefficient, regarded as a constant ;

$g$    the acceleration of the Earth's gravity ;

$N$    a coefficient of vertical eddy viscosity ;

$K, k$    coefficients of bottom friction.

Suffix $0$ denotes evaluation at the sea surface $z = 0$ and suffix $h$ evaluation at the sea bed $z = h$. In general, functional dependencies are

$$
\begin{aligned}
u, v &= u, v\,(x, y, z, t), \\
\varsigma, \bar{\varsigma}, p_a &= \varsigma, \bar{\varsigma}, p_a\,(x, y, t), \\
N &= N(x, y, z, t), \\
K &= K(x, y, t), \\
h &= h\,(x, y).
\end{aligned}
$$

A kernal function $f$ is chosen such that

$$f = f(x, y, z) \quad.$$

## Vertical integration

The equations of continuity and momentum governing the motion of the sea under wind stress, atmospheric pressure and the tide-generating forces may be written in linearized form as follows :

$$\frac{\partial \zeta}{\partial t} + \frac{\partial}{\partial x} \int_0^h u \, dz + \frac{\partial}{\partial y} \int_0^h v \, dz = 0 \quad , \tag{1}$$

$$\frac{\partial u}{\partial t} - \gamma v = - g \frac{\partial}{\partial x} (\zeta - \zeta') + \frac{\partial}{\partial z} \left( N \frac{\partial u}{\partial z} \right) , \tag{2}$$

$$\frac{\partial v}{\partial t} + \gamma u = - g \frac{\partial}{\partial y} (\zeta - \zeta') + \frac{\partial}{\partial z} \left( N \frac{\partial v}{\partial z} \right) , \tag{3}$$

where

$$\frac{\partial \zeta'}{\partial x} = \frac{\partial \zeta}{\partial x} - \frac{1}{\rho g} \frac{\partial p_a}{\partial x} \quad , \qquad \frac{\partial \zeta'}{\partial y} = \frac{\partial \zeta}{\partial y} - \frac{1}{\rho g} \frac{\partial p_a}{\partial y} \quad . \tag{4}$$

For any particular location $(x, y)$ let $f = f(x, y, z) = f(z)$ be a differentiable as yet unknown function of $z$ defined within the range $0 \leqslant z \leqslant h$. Then multiplying the momentum equations (2), (3) by $f(z)$, integrating them with respect to $z$ from $z = 0$ to $z = h$ and dividing by $h$, gives

$$\frac{\partial \hat{u}}{\partial t} - \gamma \hat{v} = - g a \frac{\partial}{\partial x} (\zeta - \zeta') + \frac{1}{h} \int_0^h f(z) \frac{\partial}{\partial z} \left( N \frac{\partial u}{\partial z} \right) dz , \tag{5}$$

$$\frac{\partial \hat{v}}{\partial t} + \gamma \hat{u} = - g a \frac{\partial}{\partial y} (\zeta - \zeta') + \frac{1}{h} \int_0^h f(z) \frac{\partial}{\partial z} \left( N \frac{\partial v}{\partial z} \right) dz , \tag{6}$$

where

$$\hat{u} = \frac{1}{h} \int_0^h f(z) u \, dz \quad , \qquad \hat{v} = \frac{1}{h} \int_0^h f(z) v \, dz \tag{7}$$

and

$$a = \frac{1}{h} \int_0^h f(z) \, dz \quad . \tag{8}$$

Treatment of the integrals in (5) and (6) is of central interest. Integrating by parts we get

$$\int_0^h f(z) \frac{\partial}{\partial z} \left( N \frac{\partial u}{\partial z} \right) dz = \left[ f(z) \cdot \left( N \frac{\partial u}{\partial z} \right) \right]_0^h - \int_0^h N f'(z) \frac{\partial u}{\partial z} \, dz , \tag{9}$$

$$\int_0^h f(z) \frac{\partial}{\partial z} \left( N \frac{\partial v}{\partial z} \right) dz = \left[ f(z) \cdot \left( N \frac{\partial v}{\partial z} \right) \right]_0^h - \int_0^h N f'(z) \frac{\partial v}{\partial z} \, dz , \tag{10}$$

where $f' = df/dz$. At this stage, wind stress conditions at the sea surface :

$$- \rho \left( N \frac{\partial u}{\partial z} \right)_0 = F_0 \quad \& \quad - \rho \left( N \frac{\partial v}{\partial z} \right)_0 = G_0 \tag{11}$$

and frictional conditions at the sea bed :

$$-\rho\left(N\frac{\partial u}{\partial \zeta}\right)_h = F_h \quad , \quad -\rho\left(N\frac{\partial v}{\partial \zeta}\right)_h = G_h \tag{12}$$

may be introduced into (9) and (10) to give

$$\int_0^h f(\zeta)\frac{\partial}{\partial \zeta}\left(N\frac{\partial u}{\partial \zeta}\right)d\zeta = f(0).\frac{F_0}{\rho} - f(h).\frac{F_h}{\rho} - \int_0^h N f'(\zeta)\frac{\partial u}{\partial \zeta}\,d\zeta \quad , \tag{13}$$

$$\int_0^h f(\zeta)\frac{\partial}{\partial \zeta}\left(N\frac{\partial v}{\partial \zeta}\right)d\zeta = f(0).\frac{G_0}{\rho} - f(h).\frac{G_h}{\rho} - \int_0^h N f'(\zeta)\frac{\partial v}{\partial \zeta}\,d\zeta \quad . \tag{14}$$

Further integration by parts on the right of (13) and (14) leads to

$$\int_0^h f(\zeta)\frac{\partial}{\partial \zeta}\left(N\frac{\partial u}{\partial \zeta}\right)d\zeta = f(0).\frac{F_0}{\rho} + N_0\,f'(0)\,u_0$$
$$-f(h).\frac{F_h}{\rho} - N_h\,f'(h)\,u_h + \int_0^h u\frac{\partial}{\partial \zeta}\{N f'(\zeta)\}\,d\zeta \quad , \tag{15}$$

$$\int_0^h f(\zeta)\frac{\partial}{\partial \zeta}\left(N\frac{\partial v}{\partial \zeta}\right)d\zeta = f(0).\frac{G_0}{\rho} + N_0\,f'(0)\,v_0$$
$$-f(h).\frac{G_h}{\rho} - N_h\,f'(h)\,v_h + \int_0^h v\frac{\partial}{\partial \zeta}\{N f'(\zeta)\}\,d\zeta \quad . \tag{16}$$

Then, taking $f$ to satisfy

$$\frac{\partial}{\partial \zeta}\{N f'(\zeta)\} = -\lambda\,f(\zeta) \tag{17}$$

where $\lambda$ is independent of $\zeta$ , it follows that

$$\int_0^h f(\zeta)\frac{\partial}{\partial \zeta}\left(N\frac{\partial u}{\partial \zeta}\right)d\zeta = f(0).\frac{F_0}{\rho} + N_0\,f'(0)\,u_0$$
$$-f(h).\frac{F_h}{\rho} - N_h\,f'(h)\,u_h - \lambda h\,\hat{u} \quad , \tag{18}$$

$$\int_0^h f(\zeta)\frac{\partial}{\partial \zeta}\left(N\frac{\partial v}{\partial \zeta}\right)d\zeta = f(0).\frac{G_0}{\rho} + N_0\,f'(0)\,v_0$$
$$-f(h).\frac{G_h}{\rho} - N_h\,f'(h)\,v_h - \lambda h\,\hat{v} \quad . \tag{19}$$

The direct substitution of (18), (19) into (5), (6) gives vertically-integrated equations :

$$\frac{\partial \hat{u}}{\partial t} + \lambda\,\hat{u} - \gamma\,\hat{v} = -ga\frac{\partial}{\partial x}(\zeta-\zeta') + X \quad , \tag{20}$$

$$\frac{\partial \hat{v}}{\partial t} + \lambda\,\hat{v} + \gamma\,\hat{u} = -ga\frac{\partial}{\partial y}(\zeta-\zeta') + Y \quad , \tag{21}$$

where

$$X = \frac{1}{\ell}\left\{ f(0).\frac{F_0}{\rho} + N_0 f'(0) u_0 - f(\ell).\frac{F_\ell}{\rho} - N_\ell f'(\ell) u_\ell \right\}, \qquad (22)$$

$$Y = \frac{1}{\ell}\left\{ f(0).\frac{G_0}{\rho} + N_0 f'(0) v_0 - f(\ell).\frac{G_\ell}{\rho} - N_\ell f'(\ell) v_\ell \right\}. \qquad (23)$$

## Eigenvalues and eigenfunctions

Having regard to dynamical boundary conditions at the sea surface $z = 0$ and at the sea bed $z = \ell$, $f$ is chosen so that $X, Y$ are conveniently reduced in form. Wind stress $F_0, G_0$ is specified at the sea surface and three cases are considered corresponding to differing frictional conditions at the sea bed. Thus

(a) with no slip at the sea bed :

$$u_\ell = v_\ell = 0 \qquad (24)$$

take

$$f(0) = 1 \;,\quad f'(0) = 0 \;,\quad f(\ell) = 0 \qquad (25)$$

so that from (22) and (23) :

$$X = F_0/\rho\ell \;,\quad Y = G_0/\rho\ell . \qquad (26)$$

(b) With linear slip at the bed :

$$F_\ell = k\rho u_\ell \;,\quad G_\ell = k\rho v_\ell \qquad (27)$$

take

$$f(0) = 1 \;,\quad f'(0) = 0 \;,\quad N_\ell f'(\ell) + k f(\ell) = 0 \qquad (28)$$

whence, again,

$$X = F_0/\rho\ell \;,\quad Y = G_0/\rho\ell . \qquad (29)$$

(c) With a general prescription of bottom stress $F_\ell, G_\ell$ take

$$f(0) = 1 \;,\quad f'(0) = 0 \;,\quad f'(\ell) = 0 \qquad (30)$$

giving

$$X = \left\{ F_0 - f(\ell) F_\ell \right\}/\rho\ell \;,\quad Y = \left\{ G_0 - f(\ell) G_\ell \right\}/\rho\ell . \qquad (31)$$

It follows that

$$\lambda = \lambda_r \;,\quad f = f_r \;,\quad r = 1, 2, \ldots, \infty \qquad (32)$$

where $\lambda_r$ denote the ascending eigenvalues and $f_r$ the corresponding eigenfunctions associated with the differential equation (17) when solved for $0 \leqslant z \leqslant h$ subject to (25) in case (a), (28) in case (b), and (30) in case (c). Writing

$$\hat{u}_r = \frac{1}{h} \int_0^h f_r(z) u \, dz \quad , \quad \hat{v}_r = \frac{1}{h} \int_0^h f_r(z) v \, dz \tag{33}$$

and

$$a_r = \frac{1}{h} \int_0^h f_r(z) \, dz \tag{34}$$

the momentum equations (20) and (21) then yield, in the respective cases,

(a)
$$\frac{\partial \hat{u}_r}{\partial t} + \lambda_r \hat{u}_r - \gamma \hat{v}_r = -g a_r \frac{\partial}{\partial x}(\zeta - \zeta') + \frac{F_0}{\rho h} \quad , \tag{35.1}$$

$$\frac{\partial \hat{v}_r}{\partial t} + \lambda_r \hat{v}_r + \gamma \hat{u}_r = -g a_r \frac{\partial}{\partial y}(\zeta - \zeta') + \frac{G_0}{\rho h} \quad , \tag{35.2}$$

(b)
$$\frac{\partial \hat{u}_r}{\partial t} + \lambda_r \hat{u}_r - \gamma \hat{v}_r = -g a_r \frac{\partial}{\partial x}(\zeta - \zeta') + \frac{F_0}{\rho h} \quad , \tag{36.1}$$

$$\frac{\partial \hat{v}_r}{\partial t} + \lambda_r \hat{v}_r + \gamma \hat{u}_r = -g a_r \frac{\partial}{\partial y}(\zeta - \zeta') + \frac{G_0}{\rho h} \quad , \tag{36.2}$$

(c)
$$\frac{\partial \hat{u}_r}{\partial t} + \lambda_r \hat{u}_r - \gamma \hat{v}_r = -g a_r \frac{\partial}{\partial x}(\zeta - \zeta') + \frac{F_0 - f_r(h) F_h}{\rho h} \quad , \tag{37.1}$$

$$\frac{\partial \hat{v}_r}{\partial t} + \lambda_r \hat{v}_r + \gamma \hat{u}_r = -g a_r \frac{\partial}{\partial y}(\zeta - \zeta') + \frac{G_0 - f_r(h) G_h}{\rho h} \quad , \tag{37.2}$$

for $r = 1, 2, \ldots, \infty$. Necessarily $f_r(h) \neq 0$ in (37.1) and (37.2) to avoid the fictitious elimination of bottom friction.

## The inverse transformation

Current components $u$, $v$ are now expressed in terms of their transforms $\hat{u}_r$, $\hat{v}_r$. This is the inverse operation to that carried out in (33).

From (17) and (32) :

$$\frac{\partial}{\partial z}\{N f_r'(z)\} = -\lambda_r f_r(z) \quad , \quad \frac{\partial}{\partial z}\{N f_s'(z)\} = -\lambda_s f_s(z) \tag{38}$$

where $r$, $s$ denote any two integers of the sequence $1, 2, \ldots, \infty$.

Hence

$$(\lambda_s - \lambda_r) f_r f_s = \frac{\partial}{\partial z}\left[ N(f_s f_r' - f_s' f_r) \right] \tag{39}$$

so that

$$(\lambda_s - \lambda_r) \int_0^h f_r f_s \, dz = N_h \left\{ f_s(h) f_r'(h) - f_s'(h) f_r(h) \right\}$$
$$- N_o \left\{ f_s(0) f_r'(0) - f_s'(0) f_r(0) \right\} . \tag{40}$$

It follows that, in each of the cases (a), (b) and (c), with (25), (28) and (30) respectively satisfied,

$$\int_0^h f_r f_s \, dz = 0 \quad (r \neq s) . \tag{41}$$

The eigenfunctions are thus orthogonal and the use of this property assuming

$$u = \sum_{r=1}^{\infty} A_r f_r(z) \quad , \quad v = \sum_{r=1}^{\infty} B_r f_r(z) \tag{42}$$

where $A_r, B_r$ are independent of $z$ , leads to the result

$$u = \sum_{r=1}^{\infty} \phi_r \hat{u}_r f_r(z) \quad , \quad v = \sum_{r=1}^{\infty} \phi_r \hat{v}_r f_r(z) \tag{43}$$

where

$$\phi_r = h \Big/ \int_0^h f_r^2 \, dz \quad . \tag{44}$$

Substitution of $u$ and $v$ from (43) into (1) gives the equation of continuity in terms of $\hat{u}_r$, $\hat{v}_r$ rather than $u$, $v$ as follows :

$$\frac{\partial \zeta}{\partial t} + \sum_{r=1}^{\infty} \left\{ \frac{\partial}{\partial x} (h a_r \phi_r \hat{u}_r) + \frac{\partial}{\partial y} (h a_r \phi_r \hat{v}_r) \right\} = 0 \quad . \tag{45}$$

The three-dimensional equations (1), (2), (3) have now been transformed into a system of two-dimensional equations involving (45) with : (35.1) and (35.2) in case (a), (36.1) and (36.2) in case (b), (37.1) and (37.2) in case (c). Solving this two-dimensional system yields $\zeta$, $\hat{u}_r$, $\hat{v}_r$ for $r = 1, 2, \ldots, \infty$ where

$$\zeta, \hat{u}_r, \hat{v}_r = \zeta, \hat{u}_r, \hat{v}_r (x, y, t) . \tag{46}$$

Solution (46) is subject to a prescribed initial condition and appropriate lateral boundary conditions along the land and open-sea sections of the sea area being modelled (Heaps 1974).

Having determined $\hat{u}_r$ and $\hat{v}_r$ in the horizontal and through time, the currents $u, v$ at any depth may be deduced from (43). Components of depth-mean current are then deducible from

$$\bar{u} = \sum_{r=1}^{\infty} \phi_r \hat{u}_r a_r \quad , \quad \bar{v} = \sum_{r=1}^{\infty} \phi_r \hat{v}_r a_r \qquad (47)$$

these expressions coming directly from (43) after integration through the depth.

## Eddy viscosity uniform through the depth

Consideration is now given to the special circumstances in which $N$ is independent of $\zeta$ . Then, writing

$$\lambda_r = N \alpha_r^2 / h^2 \quad , \qquad (48)$$

from (38) :

$$\partial^2 f_r / \partial \zeta^2 = -\alpha_r^2 f_r \quad , \qquad (49)$$

$\alpha_r$ independent of $\zeta$ , so that

$$f_r = D_r \cos \alpha_r \zeta + E_r \sin \alpha_r \zeta \qquad (50)$$

where $D_r$ , $E_r$ are constants of integration. In completing the determination of $f_r$ , cases (a), (b), (c) are considered in turn.

(a) No bottom slip

From (25) and (50) :

$$\alpha_r = (2r - 1)\pi/2 \quad , \quad f_r = \cos \alpha_r \zeta \qquad (51)$$

whence from (34) and (44) :

$$a_r = (-1)^{r+1} \alpha_r^{-1} \quad , \quad \phi_r = 2 \quad . \qquad (52)$$

With these forms, the relevant equations for solution are (45), (35.1), (35.2). As it turns out, solutions originating thus are not practically useful since a no-slip bed condition employed with a depth-uniform eddy viscosity cannot adequately represent dynamical conditions near the bed and in the vertical water column generally (Johns and Odd 1966). The use of the no-slip condition has to be accompanied by a depth-varying eddy viscosity; Davies (1977b) has solved this problem for wind-driven motion in a rectangular basin by expanding $u$ , $v$ and $N$ in terms of depth-dependent B-splines.

(b) Linear bottom slip

From (28) and (50) :

$$f_r = \cos \alpha_r \xi \tag{53}$$

where $\alpha_r$ is the $r$th positive root, in ascending order, of

$$\alpha \tan \alpha = c \quad , \quad c = K\ell/N \quad . \tag{54}$$

Then, from (34) and (44) :

$$a_r = \sin \alpha_r / \alpha_r \quad , \quad \phi_r = 2/(1 + a_r \cos \alpha_r) \quad . \tag{55}$$

Heaps (1973, 1974) and Heaps and Jones (1975, 1977) have used these forms in (45), (36.1), (36.2) to generate three-dimensional numerical solutions for wind-driven and density-driven flows in the Irish Sea, assuming that $K$ , $N/\ell$ and therefore $c$ are constants. With $c$ a constant, the roots $\alpha_r$ constitute a fixed set independent of position $x, y$ and time $t$ .

(c) Prescribed bottom stress

From (30) and (50) :

$$\alpha_r = (r-1)\pi \quad , \quad f_r = \cos \alpha_r \xi \tag{56}$$

whence from (34) and (44) :

$$a_1 = 1 \quad , \quad a_r = 0 \quad (r \geqslant 2) ,$$
$$\phi_1 = 1 \quad , \quad \phi_r = 2 \quad (r \geqslant 2) \quad . \tag{57}$$

The dynamical equations applicable in this case, namely (45), (37.1), (37.2), then reduce to

$$\frac{\partial \xi}{\partial t} + \frac{\partial}{\partial x}(\ell \hat{u}_1) + \frac{\partial}{\partial y}(\ell \hat{v}_1) = 0 \quad , \tag{58}$$

$$\frac{\partial \hat{u}_1}{\partial t} - \gamma \hat{v}_1 = -g\frac{\partial}{\partial x}(\xi - \xi') + \frac{1}{\rho\ell}(F_0 - F_\ell) \quad , \tag{59}$$

$$\frac{\partial \hat{v}_1}{\partial t} + \gamma \hat{u}_1 = -g\frac{\partial}{\partial y}(\xi - \xi') + \frac{1}{\rho\ell}(G_0 - G_\ell) \quad , \tag{60}$$

$$\frac{\partial \hat{u}_r}{\partial t} + \lambda_r \hat{u}_r - \gamma \hat{v}_r = \frac{1}{\rho\ell}\{F_0 + (-1)^r F_\ell\} \quad , \tag{61}$$

$$\frac{\partial \hat{v}_r}{\partial t} + \lambda_r \hat{v}_r + \gamma \hat{u}_r = \frac{1}{\rho\ell}\{G_0 + (-1)^r G_\ell\} \quad , \tag{62}$$

taking $r = 2, 3, \ldots, \infty$ . The expansions for $u, v$ given by (43) become simple Fourier series :

$$u = \hat{u}_1 + 2 \sum_{r=2}^{\infty} \hat{u}_r \cos\{(r-1)\pi\zeta\}$$

$$v = \hat{v}_1 + 2 \sum_{r=2}^{\infty} \hat{v}_r \cos\{(r-1)\pi\zeta\} \qquad \left.\right\} \qquad (63)$$

and from (47) :

$$\hat{u}_1 = \bar{u} \quad, \quad \hat{v}_1 = \bar{v} \quad . \qquad (64)$$

Solving (58)-(62) for $\zeta$, $\hat{u}_r$, $\hat{v}_r$ is an approach due to Clarke (1974). Any convenient form of bottom stress may be assumed. Clarke employed the quadratic law

$$F_\ell = k\rho u_\ell (u_\ell^2 + v_\ell^2)^{1/2} \quad, \quad G_\ell = k\rho v_\ell (u_\ell^2 + v_\ell^2)^{1/2} \qquad (65)$$

with $k$ a constant and

$$u_\ell = \hat{u}_1 + 2 \sum_{r=2}^{\infty} (-1)^{r+1} \hat{u}_r \quad, \quad v_\ell = \hat{v}_1 + 2 \sum_{r=2}^{\infty} (-1)^{r+1} \hat{v}_r \qquad (66)$$

from (63). Alternatively, the linear law (27) may be employed and a solution obtained corresponding to the same physical problem as in case (b). Dealing with (b) in this way is advantageous in that the frictional parameters $N$, $K$ can be chosen freely without having to satisfy (for practical convenience) a constraint involving the constancy of $c = K\ell/N$. Conceptually, there is an attractive sub-division between the depth-averaged motion generated by equations (58), (59), (60) and deviations from this motion, through the vertical, generated by equations (61), (62) for $r = 2,3, \cdots, \infty$.

Working equations

For practical purposes it is supposed that $u$, $v$ may be determined accurately enough from the sum of the first $M$ terms in each series given by (43), so that

$$u = \sum_{r=1}^{M} \phi_r \hat{u}_r f_r(\zeta) \quad, \quad v = \sum_{r=1}^{M} \phi_r \hat{v}_r f_r(\zeta) \quad . \qquad (67)$$

In cases (a) and (b) the following 2M+1 equations (coming from (45), (35) and (36)) are then solved for $\zeta$, $\hat{u}_r$, $\hat{v}_r$ :

$$\frac{\partial \zeta}{\partial t} = - \sum_{r=1}^{M} \left\{ \frac{\partial}{\partial x}(h a_r \phi_r \hat{u}_r) + \frac{\partial}{\partial y}(h a_r \phi_r \hat{v}_r) \right\} \quad, \qquad (68)$$

$$\frac{\partial \hat{u}_r}{\partial t} = -\lambda_r \hat{u}_r + \gamma \hat{v}_r - g a_r \frac{\partial}{\partial x}(\zeta - \zeta') + \frac{F_0}{\rho h} \quad, \qquad (69)$$

$$\frac{\partial \hat{v}_r}{\partial t} = -\lambda_r \hat{v}_r - \gamma \hat{u}_r - g a_r \frac{\partial}{\partial y}(\zeta - \zeta') + \frac{G_0}{\rho h} \quad, \qquad (70)$$

( $r = 1, 2, \ldots, M$ ). Numerical finite-difference solutions may be evolved using a time-stepping procedure which advances the horizontal fields of $\zeta$, $\hat{u}_r$, $\hat{v}_r$ from one time level $t$ to a later time level $t + \Delta t$ through successive intervals $\Delta t$.

Examining (67) it is evident that, in case (a), the no slip bed condition (24) is satisfied term by term since $f_r(h) = 0$ from (25). In case (b), the linear slip bed condition (27) is similarly satisfied term by term because of (12) and $\rho N_h f_r'(h) + K \rho f_r(h) = 0$ from (28). In both these cases (67) gives $(\partial u/\partial z)_0 = (\partial v/\partial z)_0 = 0$ since $f_r'(0) = 0$ from (25) and (28); it then follows from (11) that (67) yields zero surface stress, contrary to the specification of generally non-zero $F_0$, $G_0$. Such behaviour does not indicate a deficiency in the theory since, within the range $0 < z < h$, uniform convergence with increasing $M$ ensures satisfactory accuracy in determining $u$ and $v$ from (67) providing $M$ is sufficiently large. However, at and near the sea surface $z = 0$ the convergence is impractically slow. To overcome this difficulty the series for $u$ and $v$ may be completed to infinity as follows, considering eddy viscosity $N$ as uniform through the depth.

First consider case (a). From (51), (52) and (48) :

$$ a_r \to 0 \quad , \quad \lambda_r \to \infty \quad \text{as} \quad r \to \infty \tag{71} $$

so that from (69), (70) :

$$ \hat{u}_r \to \frac{F_0}{\rho h \lambda_r} = \frac{4 h F_0}{\rho N \pi^2 (2r-1)^2} \quad , \quad \hat{v}_r \to \frac{G_0}{\rho h \lambda_r} = \frac{4 h G_0}{\rho N \pi^2 (2r-1)^2} \tag{72} $$

in the same limit. Therefore, writing the infinite series for $u$ in the form :

$$ u = \sum_{r=1}^{M} \phi_r \hat{u}_r \cos\{ (2r-1)\pi z/2 \} + \Delta u \quad , \tag{73.1} $$

approximately, for sufficiently large $M$ :

$$
\begin{aligned}
\Delta u &= \sum_{r=M+1}^{\infty} \phi_r \hat{u}_r \cos\{ (2r-1)\pi z/2 \} \\
&= \frac{8 h F_0}{\rho N \pi^2} \sum_{r=M+1}^{\infty} \frac{\cos\{ (2r-1)\pi z/2 \}}{(2r-1)^2} \\
&= \frac{8 h F_0}{\rho N \pi^2} \left[ \sum_{r=1}^{\infty} \frac{\cos\{ (2r+1)\pi z/2 \}}{(2r+1)^2} - \sum_{r=1}^{M-1} \frac{\cos\{ (2r+1)\pi z/2 \}}{(2r+1)^2} \right] \\
&= \frac{8 h F_0}{\rho N \pi^2} \left[ (1-z)\frac{\pi^2}{8} - \cos\left(\frac{\pi z}{2}\right) - \sum_{r=1}^{M-1} \frac{\cos\{ (2r+1)\pi z/2 \}}{(2r+1)^2} \right] . \tag{73.2}
\end{aligned}
$$

Similarly

$$v = \sum_{r=1}^{M} \phi_r \hat{v}_r \cos\{(2r-1)\pi\zeta/2\} + \Delta v \qquad (74.1)$$

where

$$\Delta v = \frac{8 h G_0}{\rho N \pi^2}\left[(1-\zeta)\frac{\pi^2}{8} - \cos\left(\frac{\pi\zeta}{2}\right) - \sum_{r=1}^{M-1}\frac{\cos\{(2r+1)\pi\zeta/2\}}{(2r+1)^2}\right]. \qquad (74.2)$$

It may be verified that these completed expressions for $u$ and $v$ satisfy the surface stress conditions (11).

Completing to infinity the series of (67) in case (b), invoking (53)-(55) and using a similar approach to that described above, with depth-independent eddy viscosity, yields

$$u = \sum_{r=1}^{M} \phi_r \hat{u}_r \cos\alpha_r \zeta + \Delta u \qquad (75.1)$$

$$v = \sum_{r=1}^{M} \phi_r \hat{v}_r \cos\alpha_r \zeta + \Delta v \qquad (76.1)$$

where

$$\Delta u = \frac{2 h F_0}{\rho N \pi^2}\left[(3\zeta^2 - 6\zeta + 2)\frac{\pi^2}{12} - \sum_{r=1}^{M-1}\frac{\cos r\pi\zeta}{r^2}\right], \qquad (75.2)$$

$$\Delta v = \frac{2 h G_0}{\rho N \pi^2}\left[(3\zeta^2 - 6\zeta + 2)\frac{\pi^2}{12} - \sum_{r=1}^{M-1}\frac{\cos r\pi\zeta}{r^2}\right]. \qquad (76.2)$$

In case (c) the following 2M+1 equations (from (45) and (37)) are solved for $\zeta$, $\hat{u}_r$, $\hat{v}_r$ :

$$\frac{\partial \zeta}{\partial t} = -\sum_{r=1}^{M}\left\{\frac{\partial}{\partial x}(h a_r \phi_r \hat{u}_r) + \frac{\partial}{\partial y}(h a_r \phi_r \hat{v}_r)\right\}, \qquad (77)$$

$$\frac{\partial \hat{u}_r}{\partial t} = -\lambda_r \hat{u}_r + \gamma \hat{v}_r - g a_r \frac{\partial}{\partial x}(\zeta - \zeta') + \frac{F_0 - f_r(h)F_h}{\rho h}, \qquad (78)$$

$$\frac{\partial \hat{v}_r}{\partial t} = -\lambda_r \hat{v}_r - \gamma \hat{u}_r - g a_r \frac{\partial}{\partial y}(\zeta - \zeta') + \frac{G_0 - f_r(h)G_h}{\rho h}, \qquad (79)$$

( $r = 1,2, \ldots, M$ ). In this case, (67) gives

$$(\partial u/\partial\zeta)_0 = (\partial v/\partial\zeta)_0 = (\partial u/\partial\zeta)_h = (\partial v/\partial\zeta)_h = 0$$

since $f_r'(0) = f_r'(h) = 0$ from (30). Accordingly, from (11) and (12), both the surface stress and the bottom stress are zero, contrary to the general specification of non-zero $F_0$, $G_0$, $F_h$, $G_h$. Completion to infinity of the series in (67) removes this contradiction. Thus, invoking (56)-(62),

$$u = \sum_{r=1}^{M} \phi_r \hat{u}_r \cos\{(r-1)\pi\xi\} + \Delta u \qquad (80.1)$$

where approximately, for sufficiently large $M$ :

$$\Delta u = \sum_{r=M+1}^{\infty} \phi_r \hat{u}_r \cos\{(r-1)\pi\xi\}$$

$$= \sum_{r=M+1}^{\infty} \frac{2h}{\rho N \pi^2 (r-1)^2} \left[ F_0 + (-1)^r F_h \right] \cos\{(r-1)\pi\xi\}$$

$$= \frac{2h F_0}{\rho N \pi^2} \left[ \sum_{r=1}^{\infty} \frac{\cos r\pi\xi}{r^2} - \sum_{r=1}^{M-1} \frac{\cos r\pi\xi}{r^2} \right]$$

$$- \frac{2h F_h}{\rho N \pi^2} \left[ \sum_{r=1}^{\infty} (-1)^r \frac{\cos r\pi\xi}{r^2} - \sum_{r=1}^{M-1} (-1)^r \frac{\cos r\pi\xi}{r^2} \right]$$

$$= \frac{2h F_0}{\rho N \pi^2} \left[ (3\xi^2 - 6\xi + 2)\frac{\pi^2}{12} - \sum_{r=1}^{M-1} \frac{\cos r\pi\xi}{r^2} \right]$$

$$- \frac{2h F_h}{\rho N \pi^2} \left[ (3\xi^2 - 1)\frac{\pi^2}{12} - \sum_{r=1}^{M-1} (-1)^r \frac{\cos r\pi\xi}{r^2} \right] . \qquad (80.2)$$

Similarly

$$v = \sum_{r=1}^{M} \phi_r \hat{v}_r \cos\{(r-1)\pi\xi\} + \Delta v \qquad (81.1)$$

where

$$\Delta v = \frac{2h G_0}{\rho N \pi^2} \left[ (3\xi^2 - 6\xi + 2)\frac{\pi^2}{12} - \sum_{r=1}^{M-1} \frac{\cos r\pi\xi}{r^2} \right]$$

$$- \frac{2h G_h}{\rho N \pi^2} \left[ (3\xi^2 - 1)\frac{\pi^2}{12} - \sum_{r=1}^{M-1} (-1)^r \frac{\cos r\pi\xi}{r^2} \right] . \qquad (81.2)$$

These expressions for $u$ and $v$ satisfy, exactly, the surface and bottom conditions (11) and (12). First approximations to $F_h$ and $G_h$ from (65) may be obtained employing $u_h$ and $v_h$ given by (80.1) with $\Delta u = 0$ and (81.1) with $\Delta v = 0$. The approximate $F_h$, $G_h$ may then be used in (80.1) and (81.1) ( $\Delta u$, $\Delta v \neq 0$ ) to improve the estimation of $u_h$, $v_h$ and thence $F_h$, $G_h$. This procedure may be performed in each time step of a numerical solution.

## Concluding remarks

It may seem inappropriate for this paper, presented at a symposium on the modelling of estuarine and coastal physics, to have concentrated exclusively on mathematical method. However, the material discussed here needed discussion in the continuing development of a programme of three-dimensional sea modelling - mapped out retrospectively by the referenced work. Associated physical considerations and application of the mathematical method are aspects emphasised in a parallel contribution to the Proceedings of the Sixteenth Conference on Coastal Engineering.

Figures 1 and 2 show typical storm-surge elevations and currents derived by Heaps and Jones (1975) from a three-dimensional Irish Sea model based on the design principles described here.

## Acknowledgements

I am grateful to Dr. A. M. Davies for valuable comments.

The work described in this paper was funded by a Consortium consisting of the Natural Environment Research Council, the Ministry of Agriculture, Fisheries and Food, and the Departments of Industry and Energy.

## References

Clarke, D.J. 1974 Three-dimensional storm surge computations. Geophys. J. R. astr. Soc., 39, 195-199.

Davies, A.M. 1977a The numerical solution of the three-dimensional hydrodynamic equations using a B-spline representation of the vertical current profile. pp. 1-25 in, Bottom Turbulence, Proceedings of the 8th Liège Colloquium on Ocean Hydrodynamics, (ed. J.C.J. Nihoul). Amsterdam : Elsevier. (Elsevier Oceanography Series, 19).

Davies, A.M. 1977b Three-dimensional model with depth-varying eddy viscosity. pp. 27-48 in, Bottom Turbulence, Proceedings of the 8th Liège Colloquium on Ocean Hydrodynamics, (ed. J.C.J. Nihoul). Amsterdam : Elsevier. (Elsevier Oceanography Series, 19).

Davies, A.M. 1978 Application of numerical models to the computation of the wind induced circulation of the North Sea during JONSDAP '76. (In press).

Heaps, N.S. 1972 On the numerical solution of the three-dimensional hydrodynamical equations for tides and storm surges. Mém. Soc. r. sci. Liège, ser.6, 2, 143-180.

Heaps, N.S. 1973 Three-dimensional numerical model of the Irish Sea. Geophys. J. R. astr. Soc., 35, 99-120.

Heaps, N.S. 1974 Development of a three-dimensional numerical model of the Irish Sea. Rapp. P.-v. Réun. Cons. int. Explor. Mer., 167, 147-162.

Heaps, N.S. 1976 On formulating a non-linear numerical model in three dimensions for tides and storm surges. pp. 368-387 in, Computing Methods in Applied Sciences, (ed. R. Glowinski and J. L. Lions). Berlin : Springer-Verlag. (Lecture Notes in Physics, 58).

Heaps, N.S. and Jones, J.E. 1975 Storm surge computations for the Irish Sea using a three-dimensional numerical model. Mém. Soc. r. sci. Liège, ser.6, 7, 289-333.

Heaps, N.S. and Jones, J.E. 1977 Density currents in the Irish Sea. Geophys. J. R. astr. Soc., 51, 393-429.

Johns, B. and Odd, N. 1966 On the vertical structure of tidal flow in river estuaries. Geophys. J. R. astr. Soc., 12, 103-110.

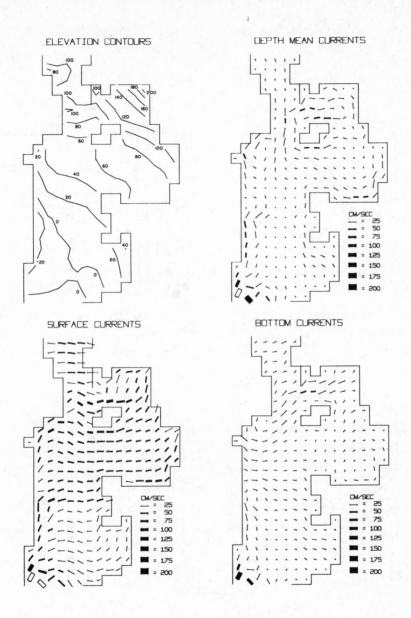

Figure 1. Meteorological effects in the Irish Sea : elevation contours (in cms) along with depth-mean currents, surface currents and bottom currents at 0000 hours on 14 January 1965 - a time of maximum surge height. From computations with a three-dimensional numerical model (Heaps and Jones 1975).

90

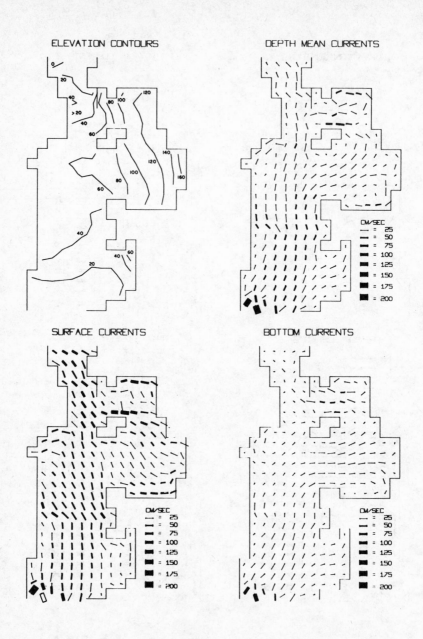

Figure 2. Meteorological effects in the Irish Sea : elevation contours (in cms) along with depth-mean currents, surface currents and bottom currents at 0400 hours on 14 January 1965 - a time when surge height was decreasing from a maximum. From computations with a three-dimensional numerical model (Heaps and Jones 1975).

# ON THE FORMATION OF SALT WEDGES IN ESTUARIES

Ernst Maier-Reimer

Institut für Meereskunde der Universität Hamburg

## Abstract

A new method to compute the transport of density in baroclinic
numerical models without artificial diffusivity is presented. As a
critical application, the salt intrusion into a schematic estuary of
the size of the Elbe estuary is studied.

The model is based on the Navier Stokes equation solved numerically
by an established method in the Eulerian framework. The transport of
density is computed in a Lagrangian is done in a convenient way using
the computed velocities. In contrast to the established tracer
particle methods, each particle represents a certain amount of water
of definite properties (salinity, temperature, etc.). The mean
property of any Eulerian grid cell is the average of all Lagrangian
particles being in the cell. Using this concept, four types of
transport can be modelled completely independent of each other:

- "large scale" transport by mean velocities
- "medium scale" displacements by turbulence
- buoyancy
- "small scale" exchange of properties by irreversible mixing.

The sharp density gradient of the brackish water zone is reproduced
qualitatively. Its dependence on the tide and on the fresh water
input can be demonstrated, too.

# 1. Introduction

For many purposes, it is convenient to construct models of estuarine
flow according to a simple typology: the one extreme is the "well
mixed" estuary in which the density depends only on horizontal
coordinates. Formally, it can be characterized by a vertical diffusion
coefficient of infinite magnitude. The other extreme is the stratified
estuary in which the vertical diffusion is zero. However, the normal
case is the "partly" mixed estuary, an equilibrium state in which
stratification and mixing balance each other. (fig. 1)

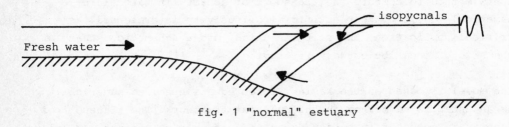

fig. 1 "normal" estuary

The arrows indicate deviations from the mean current direction which
is directed versus the open sea. These deviations involve entrainment
processes. In many estuaries, the isopycnals concentrate in a small
region, forming a brackish water zone with relatively sharp gradients.
The position of this gradient zone, its width and the slope of the
isopycnals depend strongly on the fresh water inflow and the mixing
processes which are caused mainly by the tide. At present time, the
nature of mixing processes is not clearly understood. The common way
to parametrize them by coefficients of horizontal and vertical
diffusion cannot be but a crude approximation. It may be argued that
the physically relevant coordinates should be the directions parallel
and normal to the isopycnals. The transformability onto other
coordinates requires a tensorial form of the diffusion coefficient.
On the other hand, the diffusion coefficients are no subjects of
direct measurements. The careless derivation of coefficients from
observed variances of a dye release, for instance, may be erroneous
as is indicated by the following example:
Suppose that in an idealized shear flow

$$u = \alpha z, \ w = 0 \quad ,$$

the physical diffusion process be described by

$$S_t + uS_x = DS_{zz}$$

The solution for an instantaneous dye release

$$S(x,z,0) = \delta(x)\delta(z)$$

is

$$S(x,z,t) = \frac{\sqrt{\frac{21}{4}}}{\pi\alpha Dt^2} e^{-\frac{z^2}{Dt} + \frac{3xz}{D\alpha t^2} - \frac{3x^2}{D\alpha^2 t^3}}$$

The variance $\sigma_x^2$ of $S$ in x-direction increases with $t^3$. An observer comparing simply the variance with a Fickian diffusion equation

$$\sigma_x^2 = 2 \overset{x}{D}t$$

would conclude that the diffusion coefficient $D^x$ increases with $t^2$, or, as the length scale L increases with $t^{1.5}$, that the diffusion coefficient $D^x$ increases with $L^{4/3}$, in contrast to the assumed physical value $D^x = 0$. This is, of course, an extreme example.

In real estuaries, the comparison of measurements with plausible theories (Talbot 1972, Bowden 1963) indicates diffusion rates of the order velocity times depth. Another approach to describe turbulent mixing is the definition of a diffusion velocity ranging in the order of 1 cm/s. This approach immediately shows that in many estuaries the diffusivity is small in the sense that local changes by diffusion are small compared with advective changes. In the Elbe estuary, for instance, we have to deal with quite different length scales:

| | | |
|---|---|---|
| overall length | 150 | km |
| width | 0.2 - 20 | km |
| depth | 10 | m |
| brackish zone | 20 | km |
| tidal way | 20 | km |

In this estimation, the Elbe estuary is defined as the section between the beginning of tidal signals at Geesthacht and the open shelf sea at Cuxhaven.

From the numerical point of view, the main difficulty arises from this difference in the dominant length scales. Using difference approximations in a uniform grid, the brackish zone does not cover but rather few grid points. Using the terms of a Fourier decomposition,

$$S = \sum_{\nu} a_{\nu} e^{ik_{\nu}x}$$

this means that the short waves whose wavelength is comparable to the grid size are important. Let h be the grid increment then for a wavenumber k the discretization error of a difference scheme of order n is of the order

$$\frac{kh^{n+1}}{(n+1)!}$$

The shortest wave appearing in the grid has $kh = \pi$. Consequently, when dealing with sharp gradients, an important part of the spectrum has a discretization error $\frac{\pi^{n+1}}{(n+1)!}$ . From this estimation follows that a scheme of higher order cannot always be better than a lower order scheme. It has been shown (Grotjan, O'Brien 1976) that for the advective equation in the short wave extreme a fourthorder scheme, for instance, may be even worse than the first and second order schemes.

Regarding the advection-diffusion equation,

$$S_t + u\, S_x = AS_{xx} \tag{1}$$

the critical number for difference approximation is something like a Reynolds-number, $Rd = uh/A$. Simple numerical schemes work well when $Re < 2$. This condition would require a horizontal mesh size of the order of the depth, which is not achievable. One of the most widely used schemes for eq. (1) is the upwind scheme. This is a first order scheme which inevitably involves an "artificial" diffusivity sufficiently large to fulfill the condition. (In the special case the Courant number $u\,\Delta t/h$ being 1, the upwind scheme is exact; but in practical applications, this case cannot be maintained). Several authors (e.g. Boris & Book 1973, Egan & Mahoney 1972) proposed corrections to the upwind scheme reducing drastically the artificial diffusivity. The numerical problems related to the advection equation vanish completely by a transformation on Lagrangian coordinates: the solution of $\rho_t + u\rho_x$ simply is $\rho = \rho(x - \int u dt)$. In contrast to the Eulerian framework, in the Lagrangian coordinates the diffusive part of the equation yields serious troubles (Maier-Reimer, 1973) when represented by finite differences. In this paper, another scheme is proposed which is based on a transformation on Lagrangian coordinates.

In Section 2, the numerical procedure is described, in 3, some test calculations are presented.

## 2. The Water-blob-tracing model

For an incompressible fluid, the motion can be described by the equations (neglecting external heating)

$$\frac{du_i}{dt} + \varepsilon_{ijk}f_j u_k + \frac{\partial p}{\rho \partial x_i} = \text{Reynolds stresses} \qquad (2)$$

$$\frac{\partial u_j}{\partial x_j} = 0 \qquad (3)$$

$$\rho_t + u_j \frac{\partial \rho}{\partial x_j} = \text{diffusive fluxes.} \qquad (4)$$

Knowing the distribution of $\rho$, eqs. (2) and (3) can be solved well by difference approximations in Eulerian coordinates. A description of many existing models is given, for instance, by Policastro (Policastro 1976). For the treatment of (4), let us state the Definition: the water is not a continuum, it consists of a finite number of water blobs, each blob having definite properties. The property of a Eulerian grid cell is the average property of the particles being in the cell.
The second part of the definition requires for every cell several particles. Consequently, the number of particles must be several times higher than the number of grid cells. Due to the limited computer capacity, the number cannot be chosen freely. It must be chosen high enough to ensure that empty cells are a rare exception. The abundance of empty cells follows from a combinatorial analysis: Assume M particles distributed stochastically in N cells, M > N. The mean population density is m = M/N. For an individual cell, the population probability obeys Poissons statistics when N >> 1: the probability to find n particles in the cell is $\frac{e^{-m}}{n!} m^n$,
The probability to find the cell empty is $e^{-m}$.
When N and M are kept fixed, the population densities of all cells are not independent of each other. Consequently, $1 - Ne^{-m}$ is only a crude approximation to the probability $P_{M,N}$ that all cells are non-empty. An exact analysis gives

$$P_{M,N} = \sum_{\nu=0}^{N} (-1)^{\nu} \binom{N}{\nu} \left(1 - \frac{\nu}{N}\right)^{M}$$

Assuming the density to be dependent on temperature T and salinity S, the density field then is described by M particles with geometrical and thermodynamical coordinates

$$x^{j}, y^{j}, z^{j}, T^{j}, S^{j}, \quad \rho^{j} = \rho(S^{j}, T^{j}), \quad j = 1, \ldots, M$$

The time variations of the density field result from 4 different processes independent of each other:

(A) "Large scale" transport by mean motion. This step represents the transformation on Lagrangian coordinates. Let $x^{j,i}$ be the $i^{th}$ coordinate of particle j, $m^{j,i} = [x^{j,i}/\Delta x^{i}]$ the largest integer number $\leq x^{j,i}$.

$$\text{with} \qquad \alpha^{j,i} = x^{j,i} - m^{j,i}\Delta x^{i}, \qquad \beta^{j,i} = 1 - \alpha^{j,i}$$

$$x^{j,i}(t+\Delta t) = x^{j,i}(t) + \Delta t \left[ \sum_{\nu_1, \nu_2, \nu_3 = 0}^{1} U^{i}(m^{j,1}+\nu_1, m^{j,2}+\nu_2, m^{j,3}+\nu_3) \cdot \right.$$

$$\left. \cdot \prod_{k=1}^{3} (\nu_k \alpha^{j,k} + (1-\nu_k)\beta^{j,k}) \right] \qquad (5)$$

represents the transport of particle j during a timestep $\Delta t$ by the velocity interpolated from the fixed points of the Eulerian grid. When dealing with a staggered grid, the assignment $m^{j,i} = x^{j,i}$ becomes different for the three components of velocity.

(B) Turbulent mixing. In his classical paper, Einstein derived the relation between Brownian motion and diffusion: Let f be the density distribution of any contaminant in the water, $\phi$ a displacement probability distribution during a time increment $\tau$ with

$$\int_{-\infty}^{\infty} \phi(r)\, dr = 1 \quad , \qquad \int_{-\infty}^{\infty} r\phi(r)\, dr = 0 \quad .$$

Then $\qquad f(x, t + \tau) = f(x,t) + \tau f_t(x,t) + \ldots \qquad (6)$

and $\qquad f(x, t + \tau) = \int_{-\infty}^{\infty} f(x - r, t)\phi(r)\, dr \quad .$

Assuming the relevant scales of $\phi$ to be small compared with those of f, the convolution integral can be developed into a Taylor series

$$\int_{-\infty}^{\infty} f(x-r,t)\phi(r)dr = f\int\phi dr - f_x\int r\phi dr + \frac{1}{2}f_{xx}\int r^2\phi dr + \ldots \qquad (7)$$

Comparing (6) and (7), it is obvious that f obeys Fick's diffusion equation

$$f_t = Df_{xx} \quad \text{with} \quad D = \frac{1}{2\tau}\int_{-\infty}^{\infty} r^2\phi(r)dr \quad \text{(Einstein 1905)}.$$

This mechanism can be simulated quite easily (Bugliarello 1964): Construct a probability distribution of which the variance is $2D\tau$. As D depends only on the variance of $\phi$, the simplest form is sufficient:

$$\phi(r) = \begin{cases} \frac{1}{2R} & \text{if } |r|<R \\ 0 & \text{else} \end{cases}$$

$$\int r^2\phi(r)dr = \frac{R^2}{3} \quad , \qquad R = \sqrt{6D\tau} \quad .$$

Now, add at any time step for all three coordinates displacements taken randomly from the corresponding $\phi$.

When the diffusivity D is not constant in space, Fick's diffusion equation is

$$f_t = (Df_x)_x = D f_{xx} + D_x f_x \quad .$$

In the convolution (7), $\phi$ depends on x too:

$$f(x,t+\tau) = \int_{-\infty}^{\infty} f(x-r,t)\phi(x-r,r)dr$$

$$= \int (f-rf_x + \frac{r^2}{2}f_{xx}+\ldots)(\phi-r\phi_x+\frac{r^2}{2}\phi_{xx}+\ldots)dr$$

$$= \int f\phi dr - f_x\int r\phi dr + \frac{f_{xx}}{2}\int r^2\phi dr - f\int r\phi_x dr + f_x\int r^2\phi_x dr + \frac{f}{2}\int r^2\phi_{xx}dr + \ldots$$

$$= f + \tau(Df_{xx} + 2D_x f_x + fD_{xx}) + \ldots$$

In the limit $\tau, r^2 \to 0$, $D_{xx}$ can be neglected. The term $D_x f_x$ appears twice in this representation of a variable D. This error can be cancelled by the addition of a transport velocity $-D_x$ to the mean transport (5).

(C) Buoyancy: The density of a cell is defined as the mean density of the blobs in the cell. When the particles in one cell have different densities, they are subject to buoyant forces. Assuming the buoyancy to be balanced - according to Stokes' law - by viscous forces, it can be represented by the addition of a vertical velocity $\gamma(\bar{\rho} - \rho^j)$ to (5). This step is, to a certain extent, contrary to (B): whereas (B) mixes, (C) separates the particles in a definite manner.

(D) Irreversible mixing: (B) leaves the individuality of the particles unchanged. An irreversible mixing is obtained by exchanging the properties of the blobs in any cell. Let $P^j$ be the property of particle j, $\bar{P}$ the mean property of one cell containing it. Then

$$P^j(t + \tau) = P^j(t) + \delta \cdot (\bar{P} - P^j(t))$$

for all particles represents an irreversible small scale mixing. (It is obvious that $\delta$ must not exceed 1/2).

3. A test example

Consider an idealized "2.5" - dimensional estuary in which all physical properties are assumed to be well mixed over the width of the estuary, allowing the width to be variable. Denoting by B the width, u the horizontal component of the velocity, w the vertical component, p the pressure, $\rho$ the density, the motion may be described by

$$\frac{d\ Bu}{dt} + \frac{B}{\rho} P_x = (A^v u_z)_z + (A^h (Bu)_x)_x$$

$$Bw_z + (Bu)_x = 0$$

$$P_z = \rho g$$

where $A^v$ and $A^h$ denote coefficients of eddy viscosity. As boundary conditions, at one end of the estuary the fresh-water inflow is prescribed, at the other end the tidal water elevation is prescribed.

At the surface, the kinematic boundary condition $\zeta_t + u\zeta_x = w|_{z = \zeta}$ holds.

The time variation of $\rho$ results from the WBT method of Sec. 2. As boundary conditions for $\rho$, fresh water and marine water of 1000 and 1025 kg m$^{-3}$ are prescribed. The model estuary shown in the figs. below has roughly the dimensions of the Elbe estuary between Blankenese and Helgoland; the horizontal grid size is 5 km, the vertical size is 2 m (Regarding the tidal velocities in the Elbe, this grid size would require a diffusion coefficient of 3000 m$^2$/s for Rd < 2). At the fresh water end, a one-dimensional canal representing the upper part of the Elbe river up to the end of tidal phenomena is added. This canal is not shown in the figs.

As initial condition for $\rho$, a constant horizontal gradient is prescribed which is the worst initial condition for obtaining sharp gradients. The initial population density is chosen according to the width; starting with 4 blobs per cell at the upper end, the number increases until up to 60 blobs per cell at the lower end. The total number of particles is 7200. In order to ensure the maintaining of the boundary conditions, the distribution regime of particles is chosen greater than the estuary under consideration, especially at the lower end.

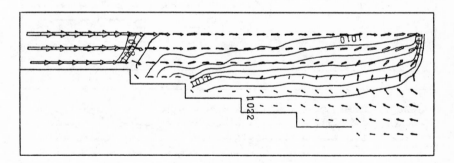

Fig. 2: Current vectors and isopycnals in the
model estuary without tide. The weakness
of the mixing is not fully consistent with
the outer boundary condition.

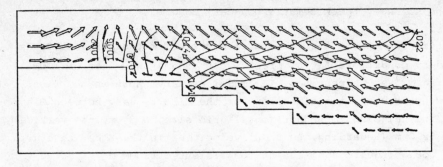

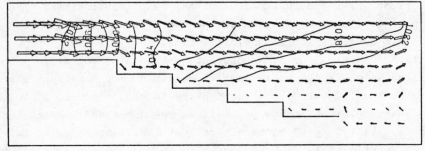

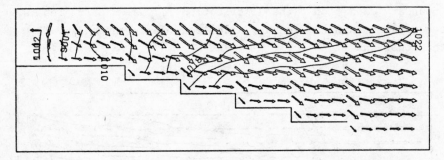

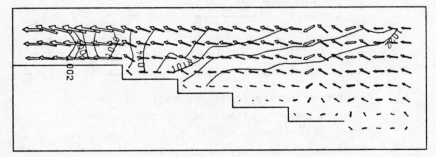

Fig. 3: Current vectors and isopycnals in the
model estuary at four states of the tidal
cycle. The model parameters are the same
as in Fig. 2.

REFERENCES


Boris, J.P. and Book, D.L. 'Flux-Corrected Transport, I. Shasta,
A Fluid Transport Algorithm that works', J. Comp. Ph. 10, 48-70 (1973)

Bowden, K.F. 'The mixing processes in a tidal estuary', Journal
of Air Water Pollution, 7 (1963) 343-356.

Bugliarello, G. and Jackson, E.D. 'Random walk study of convective
diffusion', ASCE Journal of the Engineer. Mechanics Division, 94
(1964) 49-77.

Egan, Bruce A. and Mahoney, James R. 'Numerical modelling of advection
and diffusion of urban area source pollutants', J. Appl. Met. 11,
312-322, (1972).

Einstein, A. 'Über die von der molekularkinetischen Theorie der Wärme
geforderte Bewegung von in ruhenden Flüssigkeiten suspendierten Teil-
chen', Annalen der Physik IV. (1905) 549-560.

Grotjahn, R. and O'Brien, J.J. 'Some Inaccuracies in Finite
Differencing Hyperbolic Equations', Monthly Weather Review 104 (1976)
180-194.

Maier-Reimer, E. 'Hydrodynamisch-numerische Untersuchungen zu hori-
zontalen Ausbreitungs- und Transportvorgängen in der Nordsee', Mitt.
d. Inst. f. Meereskunde d. Univ. Hamburg Nr. 21 (1973).

Policastro, A.J. and Dunn, W.E. 'Numerical modelling of surface
thermal plumes'International summer course "Heat disposal from power
generation", Dubrovnik (1976).

Talbot, J.W. 'The influence of tides, waves and other factors on
diffusion rates in marine and coastal situations', Marine pollution
and sea life, West Byfleet (1972).

## ON CURRENTS IN THE GERMAN BIGHT
A three-dimensional non-linear tidal model

J. Backhaus
Deutsches Hydrographisches Institut
Hamburg

## Abstract

A three-dimensional non-linear tidal model of a shallow coastal sea
is described. The simulation of the tide and wind generated current
regime in the German Bight, in comparison with some observations,
provides information about the physical processes which are dominant
in that part of the North Sea.

## Introduction

In the German Bight - a small part of the southeastern North Sea -
there is extensive shipping, fishing, and dumping of industrial and
municipal wastes; on the other hand, the seaside is known to be a
splendid recreation area of man. In the largest area of tidal flats
in the North Sea, along the coast of the German Bight, about 70 % of
the total North Sea fish spawn and develop. Owing to the fact that
shipping, dumping of wastes, and the number of oil production platforms, pipelines etc. show a tendency to increase, the natural equilibrium of the German Bight's ecosystem might be in danger.
Obviously, there are sufficient reasons for an extensive 'status quo
ante' survey of the relevant physical, biological, and chemical
parameters, and that is to be carried out within the framework of the
1979 "YEAR OF GERMAN BIGHT" and "MARSEN I" experiments.
The field experiments will be accompanied by simulation runs of the
three-dimensional model, here described. A previous step in the development of an appropriate model version for the dynamics in the
German Bight was presented at the 10th International Colloquium on
Ocean Hydrodynamics in Liège and will be published in the Proceedings
of that meeting (Backhaus, 1979). That model version will be referred
to as 'model A' in this text.
The significant difference between the two model versions, besides a
slight reduction in the area covered by this model (Fig. 1), is a
finer vertical discretisation and the incorporation of the non-linear
terms in the equations of motion. The horizontal resolution of 3

nautical miles was left unchanged. The reason for a better vertical
resolution of the model, is that model A was not able to resolve the
vertical current profile in a satisfactory manner, especially for
the near-surface region. The surface elevations (already described
for model A) are simulated correctly by both models, so that this
study will be related mainly to the simulation of currents and their
vertical profiles.

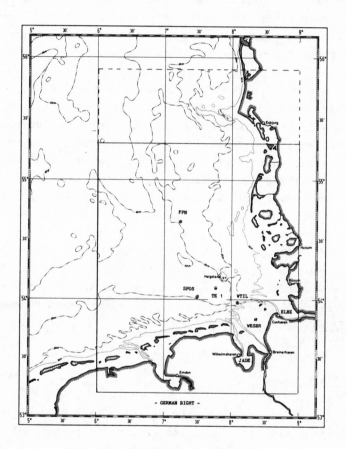

Fig. 1, German Bight, area covered by model indicated by solid
        line (dashed line for model A).

There is a demand for a fine mesh, non-linear, three-dimensional model, resulting from the dynamical situation to be found in the German
Bight. This part of the North Sea is rather shallow (maximum depth:
50 m) and, therefore, topography has a strong influence upon the
dynamics. A rather complicated system of drying banks and little islands along the shallow coast, and remarkable depth gradients, oc-

curring also in the deeper areas, require a fine horizontal grid-
resolution. The non-linear terms in the equations of motion have been
incorporated, because dynamics in a shallow tidal regime are expected to be highly non-linear.

Observations show a strong vertical current shear, especially in
near-surface regions. Some significant types of current profiles are
known, which could be viewed as being representative for certain
areas in the German Bight. The different behaviour of the currents
indicate the strong influence of the bathymetry on their vertical
distribution. These features cannot be explained by the density
structure.
Generally, in most parts of the area covered by the model, there are
counter-clockwise tidal currents over the entire water column. An
exception occurs in the ELBE/WESER estuaries, where the currents
rotate clockwise at the surface, and counter-clockwise near the bottom. In those parts, where depths are deeper than about 30 m, the
surface current ellipses are distorted to a nearly alternating current system (showing variable senses of rotation); whereas, in the
deeper regions, left rotating tidal currents are predominant. This
indicates a comparatively strong vertical current shear. In addition, phase shifts of up to one hour between surface and bottom currents are observed. In areas with depths of less than about 25 to
20 m, there is not much differences between the shape of the top and
the bottom current figures (current ellipses), indicating zones of
minimal shear.

## Description of the model

The model is designed to simulate the dynamics in a well mixed coastal sea. The assumption of well mixed conditions is not valid during
the summer - when some stratification can be observed - but there is
strong evidence that baroclinic effects are at least one order of
magnitude smaller than those arising from bottom turbulence and non-
linear wind-tide interaction. Under those circumstances, a barotropic
model version seems to be justified. The governing equations, the
list of symbols and parameters, and further details about the model
are given below. The equations of motion (1) are written in momentum
form. The sea surface displacement $\zeta$ is obtained from the vertical
velocity component at the surface, computed from the equation of

continuity (2).

$$U_t + (U^2/h)_x + (VU/h)_y + (WU/h)_z - fV + gh\,\zeta_x =$$

$$(A_H U_x)_x + (A_H U_y)_y + (A_V U_z)_z + hF^x \tag{1}$$

$$V_t + (UV/h)_x + (V^2/h)_y + (WV/h)_z + fU + gh\,\zeta_y =$$

$$(A_H V_x)_x + (A_H V_y)_y + (A_V V_z)_z + hF^y$$

$$U_x + V_y + W_z = 0 \quad ; \quad \zeta_t = W_1/h_1 \tag{2}$$

$$\tau_B = r\,\underline{MH}\,|\underline{MH}|\,/h_B^2 \quad \text{bottom stress } (h_B = \text{thickness of bottom layer})$$

$$\tau_S = c_D\,\underline{SW}\,|\underline{SW}| \qquad \text{wind induced stress at sea surface}$$

Where $\underline{M}$, $\underline{MH}$ = total, and horizontal transport (per layer), respectively; $\underline{M}$ = $(U,V,W)$; $\zeta$ = surface elevation; h = layer thickness, D = undisturbed water depth, actual water depth : H = D + $\zeta$; g = acceleration due to gravity, f = coriolis parameter; surface drag coefficient $c_D = 3.2*10^{-6}$; bottom friction parameter r = $2.5*10^{-3}$; $\underline{SW}$ = surface wind; vertical eddy viscosity coefficient $A_V$ = 50 cm$^2$s$^{-1}$; depth dependent horizontal exchange coefficient $A_H$ = h x 5 ms$^{-1}$; $F^{x,y}$ = external acceleration.
Grid spacing 3 nautical miles, time step = 169.4 s, depth of interfaces: 0, 6, 14, 24, 50 m, thickness of layers: 6, 8, 10, max. 26 m.
x,y,z = coordinates (positive east, north, and down respectively).

The system of equations is solved by applying the following boundary conditions.

At closed boundaries there is no flow normal to the wall; for parallel flow, a slip condition is applied. At open boundaries surface elevations for a progressive (barotropic) tidal wave are prescribed (in this case previously computed for tide and/or wind situations by means of a vertically integrated North Sea model). For the computation of spatial derivatives of second order, it was assumed that the first derivative of the flow normal to the open boundary vanishes.

At surface and bottom, stresses are parameterized by quadratic friction laws. The value for the non-dimensional surface drag coefficient was chosen to be the same as in the vertically integrated storm-surge models; and therefore, it might be somewhat overestimated for a three-dimensional model. However, for example, as proposed by Simons (1975) for the Great Lakes, the magnitude of this parameter must still be determined by means of a consistent and synoptic set of wind, current, water level, and sea state data, which is not yet available. This is one of the major purposes of the 1979 oceanographic experiments in the German Bight.

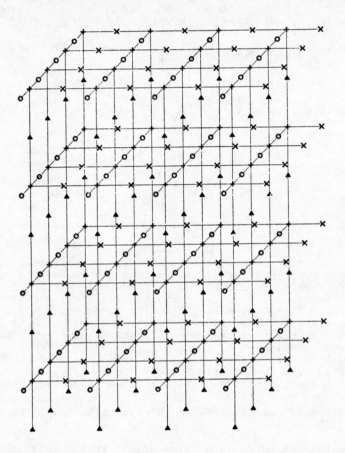

Fig. 2, Three-dimensional staggered grid, notation for grid points:
ⅹ U, ● V, ▲ W, + ζ (pressure).

The system of partial differential equations is solved approximately
by means of an explicit numerical difference scheme, which is formulated for spatially and temporally staggered grids (Sündermann, 1971).
For the three-dimensional (staggered) arrangement of grid points, see
Fig. 2.

The current profiles are approximated in the model by a vertical discretization, using five horizontal computation levels (Fig. 3), yielding four (layerwise vertically integrated) current vectors at each
horizontal grid point. The vertical resolution was chosen to be finest
for the near-surface region, where the largest insufficiencies of
model A occurred.

The internal shear stresses τ are parameterized by a constant vertical eddy viscosity coefficient. The magnitude of this parameter was
determined by means of a model-experiment. In order to find a value
which is more or less representative for the German Bight, the effect
of the topography upon the current profiles had to be eliminated. A
flat-bottomed, open box model was constructed, in order to study the
influence of the vertical eddy viscosity coefficient on the profile

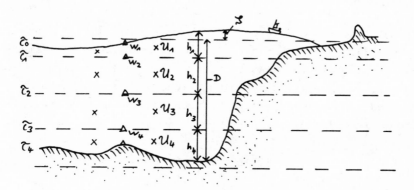

Fig. 3, Arrangement of computation levels (x-z section)

of tidal currents having the same magnitude as those in the German
Bight. The open box model was driven by an artificial boundary condition (surface elevations), taken from the analytical solution of a
propagating Kelvin wave. In order to check this test model, it was
run without non-linear terms and omitting bottom and internal friction. The results were very close to the currents and surface elevations given by the analytical solution of the Kelvin wave.
The amplitude of the surface elevation, the uniform water depth, and
the area covered by the model, were then adjusted to be similar to a
small, flat-bottom region in the German Bight, where disturbances
caused by the influence of topography were assumed to be negligible.

Various runs for different eddy viscosity coefficients were carried
out, with non-linear terms and bottom friction included. By introducing friction, the currents became rotating, and a vertical profile
developed. The current profiles during one tidal cycle, obtained in
the centre of the open box test-model, were compared with profiles
measured in that particular area in the German Bight which was assumed
to be representative. It was found that a vertical eddy viscosity
coefficient of 50 $cm^2 s^{-1}$ gave the best approximation. This value was
taken for all computations carried out with the German Bight model.

The author is well aware that the artificial boundary condition, used for the open box model, is by no means consistent with a problem including friction. Nevertheless, the deviation of surface elevations, between the analytical solution and the numerical model, was less than 2 cm at tidal range of about 2.5 m.

## Application of the model

Similar to model A, all boundary values (surface elevations) for the German Bight model were computed by previous runs with a two-dimensional North Sea model. This was carried out for all computations for the M2-tide alone, as well as for simulations with a superimposed windfield. Wind-induced changes in the mean sea level, therefore, are incorporated in the boundary values.

## Comparison of measured and computed tidal currents

A comparison with computed currents is shown for some locations in the German Bight (see chart in Fig. 1), where measurements at different depths are available. Numerous current measurements have been carried out in the past, but only a very few were made at different depths during a deployment period, so that there is a lack of locally synoptic measurements. Some of the observations shown (Figs. 4c, d) were carried out in 1924 by Thorade; he used Ekman-Merz and pendulum current meters. In comparison with the model results, his measurements appear to be very similar. This is remarkable, because he seemed to be in doubt about the accuracy of his current meters, which indeed, look quite strange when compared with modern devices. Nevertheless, he was the first who reported the significant behaviour of currents in the ELBE/WESER estuaries: a change in the sense of rotation from surface to bottom, which is also simulated by the model (see Station WTIL, Fig. 4d). At Station TE 1 (Fig. 4b), there is a change in the sense of rotation in the near-surface measurements at spring tide, which is not reproduced by the model because it runs for the $M_2$-tide only. For Station FPN (research platform "NORDSEE") no pronounced difference between top and bottom currents are observed (Fig. 4a). As far as can be judged to date, the known features of the different tidal current profiles are approximated by the model in a satisfactory manner.

The_wind_and_tide_induced_residual_circulation

All residual current situations shown were computed by integrating
the model results over one tidal cycle, when the model reached a
quasi steady state, which happens after at least five to six cycles.
The strong influence of the wind on the near-surface currents and on
the vertical distribution of the residual currents becomes evident,
for example, from the results at Station FPN (see Figs. 5 and 4a),
and at Station WTIL (Fig. 4d).

The non-linear character of the equations of motion leads to the development of some eddies in the tidal residuals for the inner German
Bight (Fig. 6), which was not simulated by model A.
The residual current distribution in the German Bight for some
(idealized) homogenous wind situations over the entire North Sea,
computed for a moderate wind speed of 5 m/s, are shown for winds
coming from North-West, South-West and North-East (Figs. 7, 8, 9).
Note that the residuals shown in those Figures are plotted in a different scale than that in Fig. 6. As far as there are measured residual currents in different depths available, the agreement between
model results and observations is fairly good. A comparison of currents in the surface layer (0 to 6 m) cannot be carried out, because
- for technical reasons - there are no reliable measurements available which are closer to the sea surface than about 8 m. There is
hope that this "verification gap" will be overcome by next year's
MARSEN I experiment, when extensive measurements of near-surface
currents, by means of modern HF techniques, will be carried out.

As already pointed out in the model A study, the residual currents in
the first metres (0 to 15 m) of the water column, where the strongest
vertical shear occurs, are depending much more upon the direction of
the wind than the near-bottom currents. Residuals in near-bottom
regions are mainly dependent upon the mean surface slope, and they
are far more persistent than the near-surface currents. All westerly
winds cause a rise of the mean water level in the German Bight which
leads to an outward directed compensation flow in the deeper layers,
focused by the 30 m depth contour. For easterly wind directions, conditions are roughly the opposite.
The change from the near-surface current system to the current regime
in the bottom layers is rather sharp and pronounced. There is evidence, from the results of both models, that the depth of the transi-

tion zone is largely unaffected by the wind speed, but affected mainly by the wind direction (see Fig. 5). Together with these flow patterns, upwelling and downwelling (depending upon the wind situation)
of watermasses is computed. These features are very pronounced in the
area South-West of the Island of Heligoland, around Station TE 1. It
has already been pointed out by Goedecke (1968) that the most intensive mixing of watermasses in the German Bight takes place around
Heligoland.

Concluding remarks

The improved model version described here permits a more accurate
simulation of the dynamical processes in a shallow coastal sea than
was possible with model A.
In comparison with some observations, the results already provide an
impression about the rather complex structure of currents and residual currents in the German Bight. However, a reliable verification
of the model is still necessary, and will be carried out by means
of the synoptic field data to be obtained during the 1979 oceano-
graphical experiments in the German Bight.

References

Backhaus, J., 1979. First Results of a Three-Dimensional Model on the
     Dynamics in the German Bight. In press (Elsevier).
Goedecke, E., 1968. Über die hydrographische Struktur der Deutschen
     Bucht im Hinblick auf die Verschmutzung in der Konvergenzzone.
     Helgoländer wiss. Meeresunters. 17, 108-125.
Simons, T.J., 1975. Effective Wind Stress over the Great Lakes
     Derived from Long-Term Numerical Model Simulations. Atmosphere,
     Vol. 13, no 4.
Sündermann, J., 1971. Die hydrodynamisch-numerische Berechnung der
     Vertikalstruktur von Bewegungsvorgängen in Kanälen und Becken.
     Mitt. Inst. f. Meereskunde, XIX.
Thorade, H., 1928. Gezeitenuntersuchungen in der Deutschen Bucht.
     Archiv der Deutschen Seewarte 46, Nr. 3.

15016    FPN

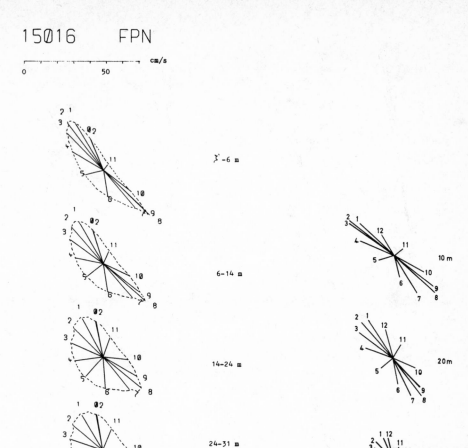

Fig. 4a, Computed (left) and observed (right) tidal currents at
research platform "Nordsee", station FPN.

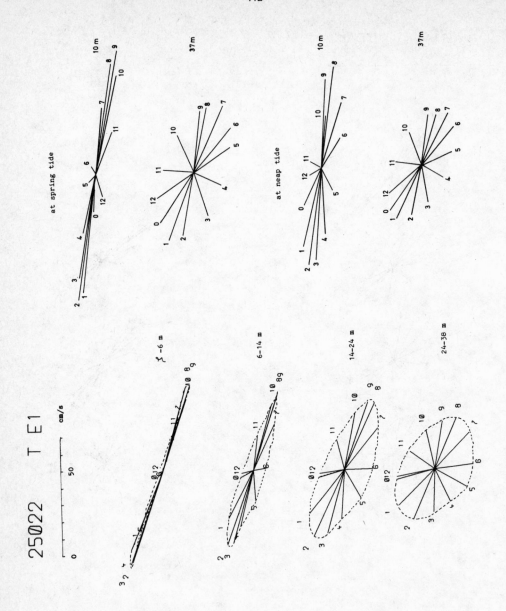

Fig. 4b, Computed (left) and observed (right) tidal currents at station TE 1.

113

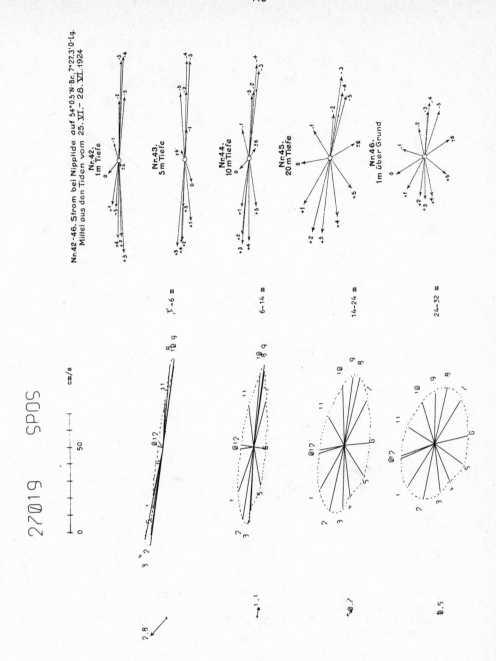

Fig. 4c, Computed (left) and observed (right) tidal currents at station SPOS.

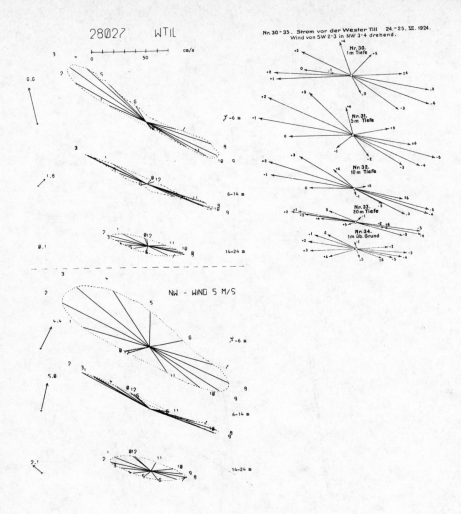

Fig. 4d, Computed (left) and observed (right) currents for tide, and tide plus wind at station WTIL. Magnitude (cm/s) and direction of computed residual currents indicated by arrows (left).

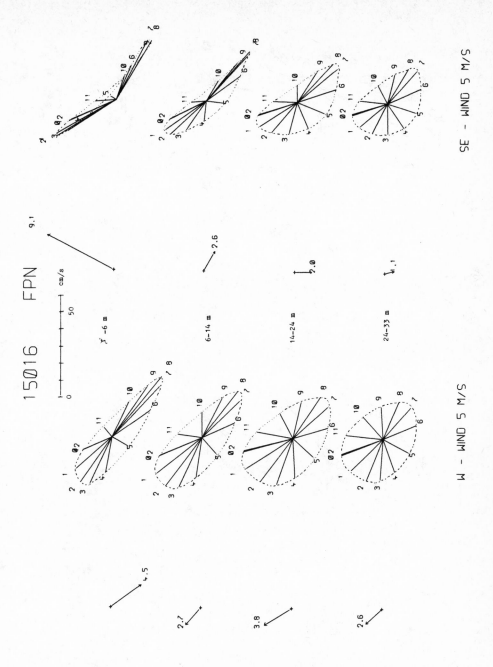

Fig. 5a, Computed wind and tide induced currents at station FPN,
magnitude (cm/s) and direction of residual currents
indicated by arrows (left).

116

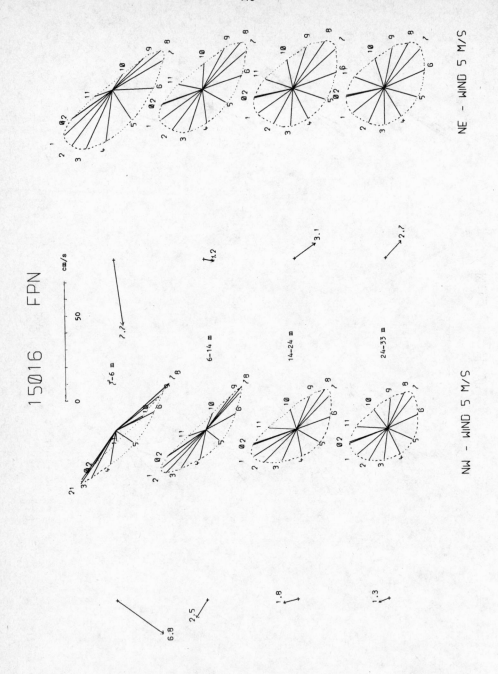

Fig. 5b, Computed wind and tide induced currents at station FPN,
magnitude (cm/s) and direction of residual currents
indicated by arrows (left).

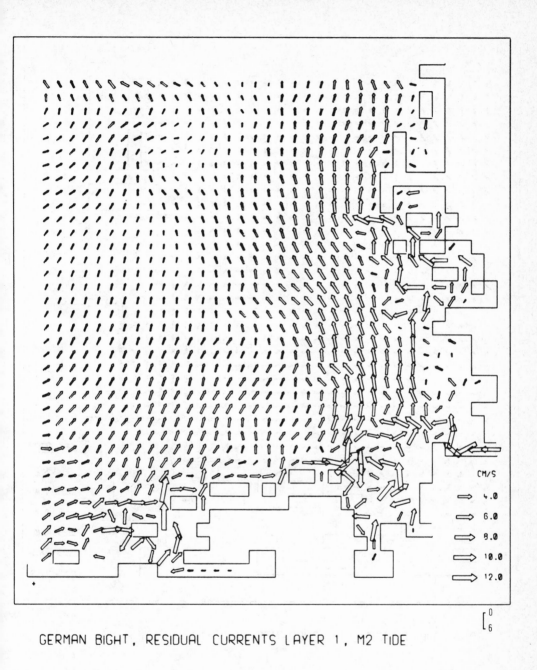

GERMAN BIGHT, RESIDUAL CURRENTS LAYER 1, M2 TIDE

Fig. 6    Vertical distribution of computed horizontal tidal residual currents ($M_2$ tide).

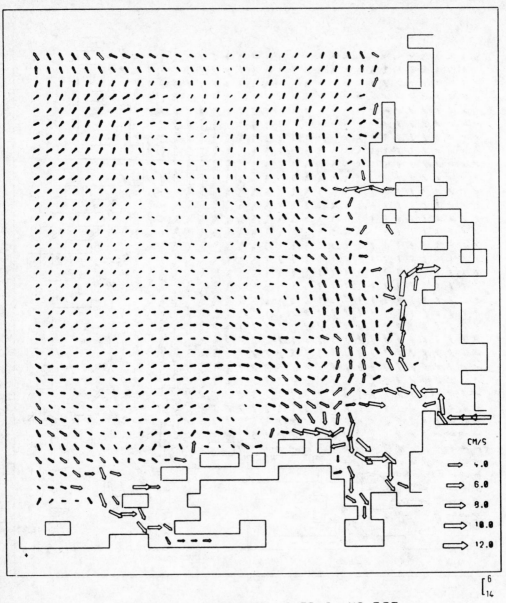

GERMAN BIGHT, RESIDUAL CURRENTS LAYER 2, M2 TIDE

Fig. 6    (ctd.)

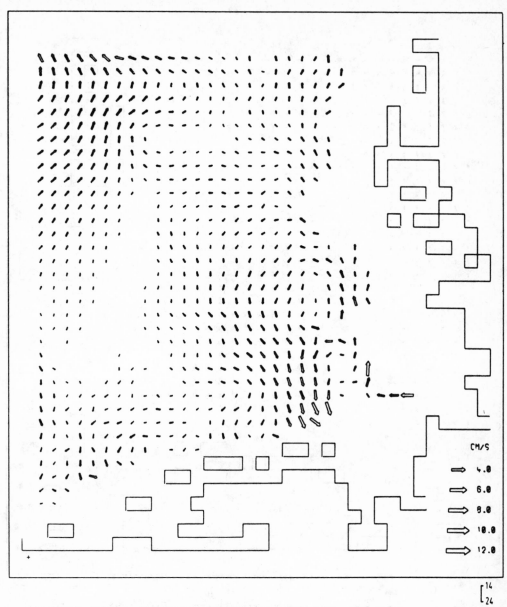

GERMAN BIGHT, RESIDUAL CURRENTS LAYER 3, M2 TIDE

**Fig. 6**   (ctd.)

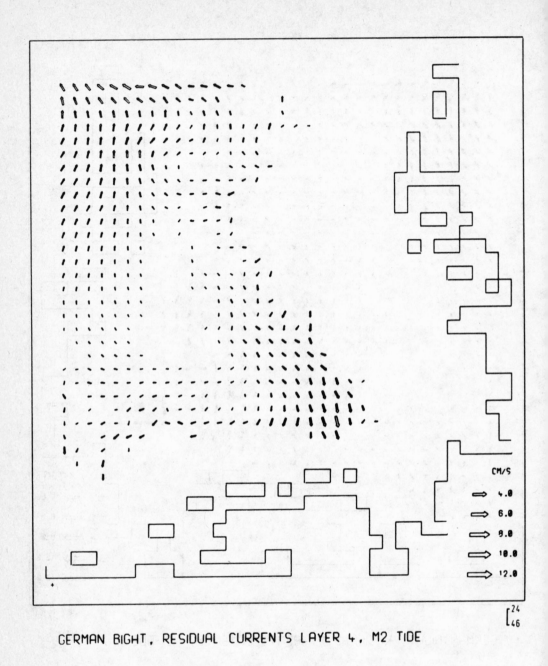

GERMAN BIGHT, RESIDUAL CURRENTS LAYER 4, M2 TIDE

Fig. 6    (ctd.)

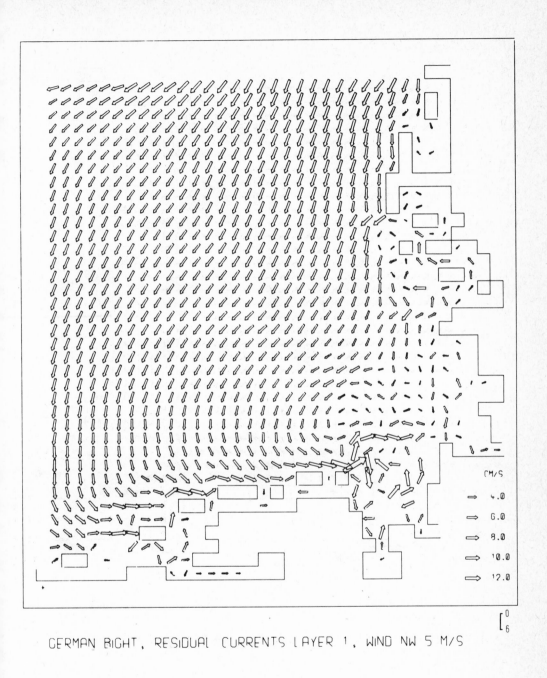

GERMAN BIGHT, RESIDUAL CURRENTS LAYER 1, WIND NW 5 M/S

Fig. 7   Vertical distribution of computed horizontal
         tidal and wind induced residual currents
         (NW wind, 5 m/s).

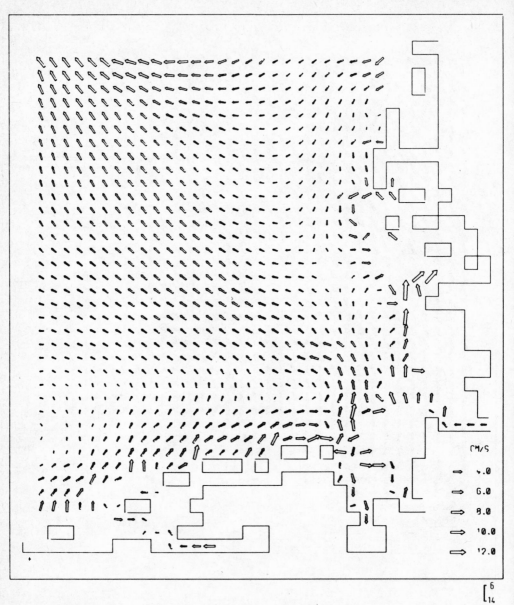

GERMAN BIGHT, RESIDUAL CURRENTS LAYER 2, WIND NW 5 M/S

Fig. 7 (ctd.)

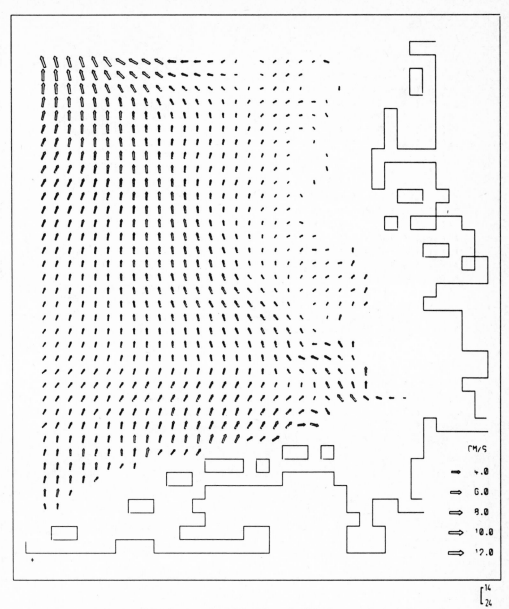

GERMAN BIGHT, RESIDUAL CURRENTS LAYER 3, WIND NW 5 M/S

Fig. 7   (ctd.)

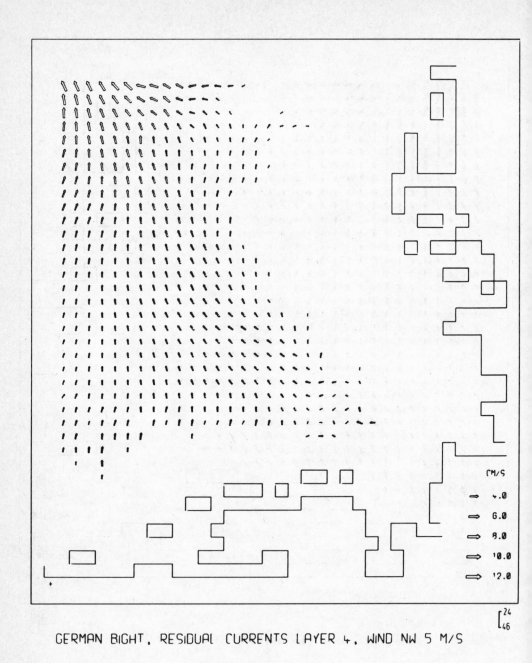

GERMAN BIGHT, RESIDUAL CURRENTS LAYER 4, WIND NW 5 M/S

Fig. 7   (ctd.)

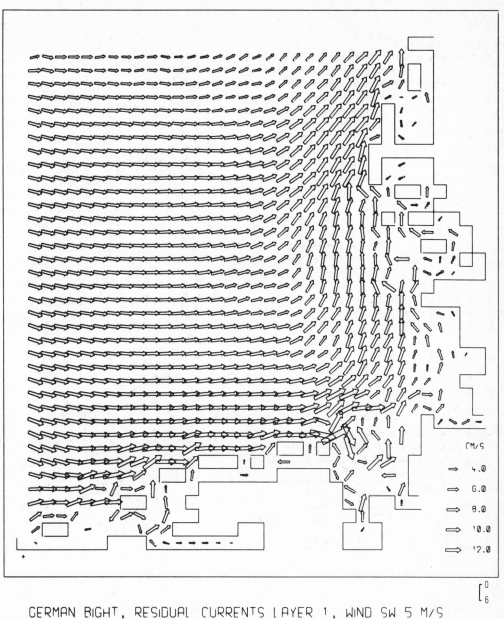

CM/S

→ 4.0

→ 6.0

→ 8.0

→ 10.0

→ 12.0

GERMAN BIGHT, RESIDUAL CURRENTS LAYER 1, WIND SW 5 M/S

Fig. 8    Vertical distribution of computed horizontal
          tidal and wind induced residual currents
          (SW wind, 5 m/s).

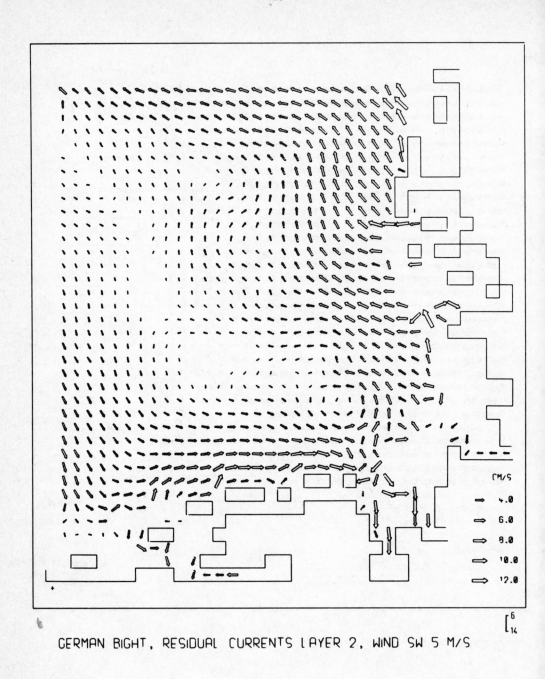

GERMAN BIGHT, RESIDUAL CURRENTS LAYER 2, WIND SW 5 M/S

Fig. 8   (ctd.)

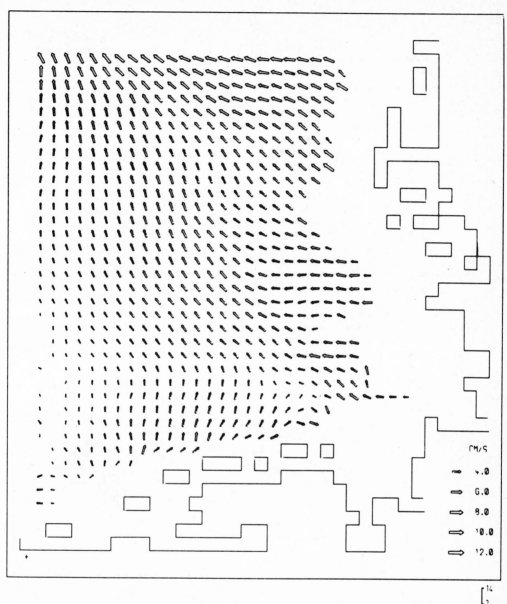

CM/S

➡ 4.0

➡ 6.0

➡ 8.0

➡ 10.0

➡ 12.0

GERMAN BIGHT, RESIDUAL CURRENTS LAYER 3, WIND SW 5 M/S

Fig. 8    (ctd.)

128

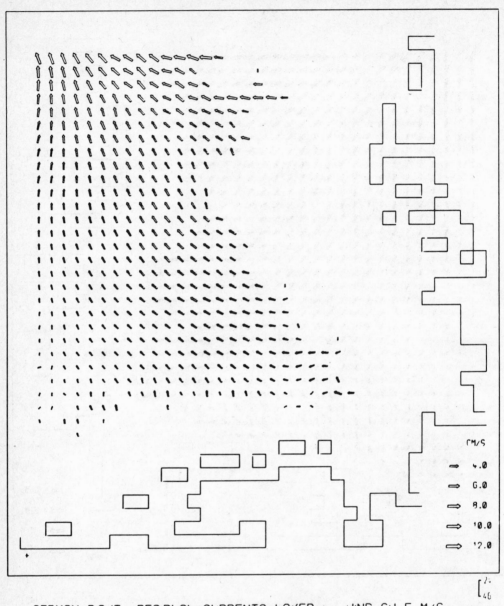

GERMAN BIGHT, RESIDUAL CURRENTS LAYER 4, WIND SW 5 M/S

Fig. 8    (ctd.)

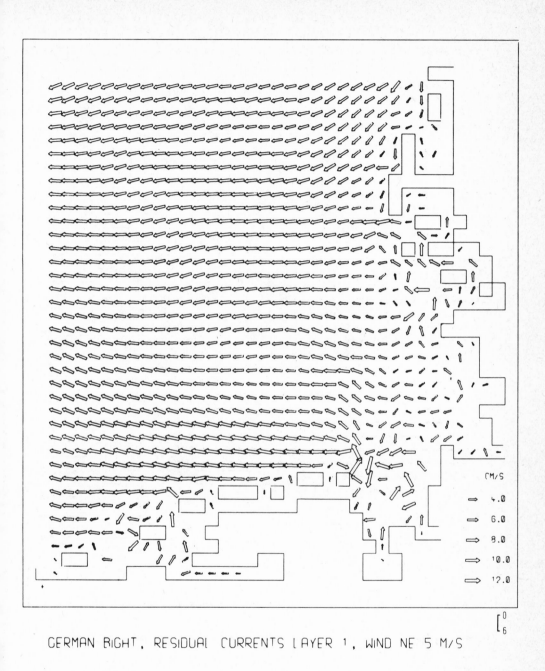

GERMAN BIGHT, RESIDUAL CURRENTS LAYER 1, WIND NE 5 M/S

**Fig. 9**  Vertical distribution of computed horizontal
tidal and wind induced residual currents
(NE wind, 5 m/s).

130

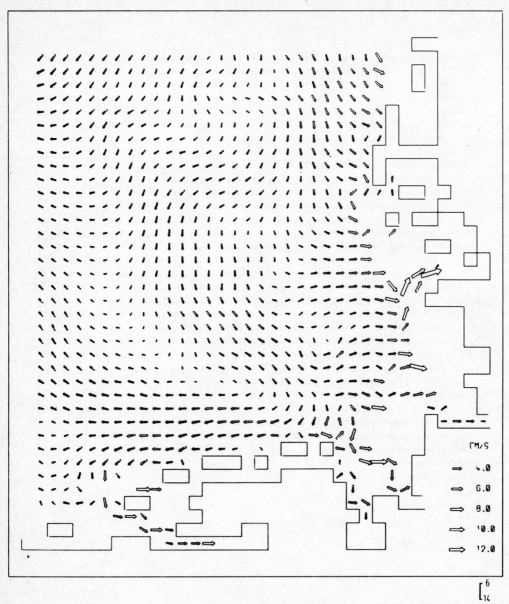

GERMAN BIGHT, RESIDUAL CURRENTS LAYER 2, WIND NE 5 M/S

Fig. 9    (ctd.)

131

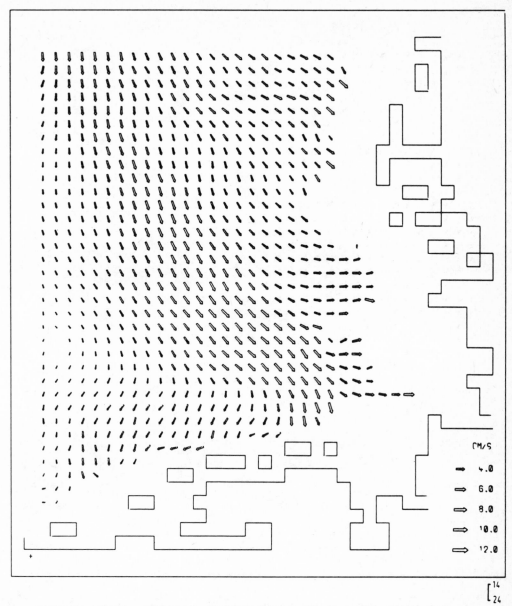

GERMAN BIGHT, RESIDUAL CURRENTS LAYER 3, WIND NE 5 M/S

Fig. 9    (ctd.)

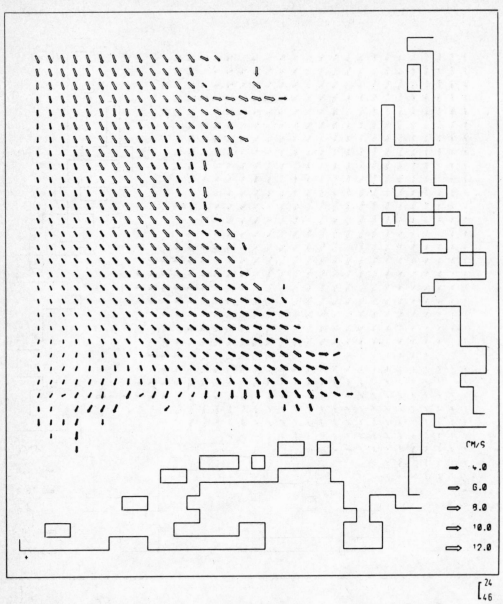

GERMAN BIGHT, RESIDUAL CURRENTS LAYER 4, WIND NE 5 M/S

Fig. 9   (ctd.)

# TIDE-INDUCED RESIDUAL FLOW

J. van de Kreeke and S. S. Chiu

Division of Ocean Engineering
Rosenstiel School of Marine and Atmospheric Science
University of Miami
Miami, Florida USA

## Objectives

The objective of this paper is to demonstrate the mechanisms by which tide-induced residual flow is generated in shallow bays and channels.

## Definitions

In the mathematical formulation of the problem use is made of a Cartesian coordinate system. The origin of the system is at the Still Water Level and the z-axis is positive upward.

The horizontal Eulerian velocities in the x and y direction are respectively u and v. u and v are assumed to be independent of z. Furthermore

$$u = u_E + u_p \tag{1a}$$

$$v = v_E + v_p \tag{1b}$$

in which $u_E$ and $v_E$ are the Eulerian residual or mean velocities, obtained by time averaging the Eulerian velocities, and $u_p$ and $v_p$ are the tidal periodic components of the Eulerian velocity.

The flow in the x and y direction is defined as

$$q_x = (h + \eta)\, u \tag{2b}$$

$$q_y = (h + \eta)\, v \tag{2b}$$

in which $h(x, y)$ is depth with respect to the Still Water Level and $\eta(x, y, t)$ is the water level with respect to the Still Water Level. Similar to the velocities

$$\eta = \bar{\eta} + \eta_p \tag{3}$$

The residual flow follows from time averaging the instantaneous flows

$$\bar{q}_x = \overline{(h + \eta)\, u} = h\, u_E + \overline{\eta_p u_p} \tag{4a}$$

$$\bar{q}_y = \overline{(h + \eta)\, v} = h\, v_E + \overline{\eta_p v_p} \tag{4b}$$

In certain problems the Lagrangian mean velocity defined as the tidally averaged velocity of a drogue is of importance. As shown by Longuet-Higgins (1969) for shallow water waves and assuming $u_E \ll u_p$ and $v_E \ll v_p$, the Lagrangian mean velocity and the residual flow are related by the approximate expressions

$$u_L = \frac{\overline{q}_x}{h} + \frac{1}{h} \frac{\partial}{\partial y} \overline{hu_p \int v_p \, dt} \tag{5a}$$

$$v_L = \frac{\overline{q}_y}{h} + \frac{1}{h} \frac{\partial}{\partial x} \overline{hv_p \int u_p \, dt} \tag{5b}$$

$u_L$ and $v_L$ are the Lagrangian mean velocities in respectively the x and y direction. Note that in contrast to flow in a channel (one-dimensional flow), for shallow seas (two-dimensional flow) the Lagrangian mean velocity is no longer the equivalent of the residual flow divided by the depth. For two-dimensional flow the integral terms cannot be neglected apriori e.g. as assumed in Nihoul and Ronday (1975) and Tee (1976).

## Residual Flow in a Sea-level Canal

To illustrate the nature of the residual flow consider the tidal motion in a relatively shallow sea-level canal. The governing one-dimensional equations are

$$\frac{\partial \eta}{\partial t} + \frac{\partial q}{\partial x} = 0 \tag{6}$$

$$\frac{\partial q}{\partial t} + \frac{2}{h} q \frac{\partial q}{\partial x} + g \, (h + \eta) \frac{\partial \eta}{\partial x} = - \frac{F_\ell q}{h+\eta} \tag{7}$$

Here the subscript x in $q_x$ has been omitted.

Approximate solutions for the unknowns q and η in Eqs. (6) and (7) are derived by writing

$$q = \overline{q} + q_1 \, (x,t) + q_2 \, (x,t) + \ldots \tag{8}$$

$$\eta = \overline{\eta} + \eta_1 \, (x,t) + \eta_2 \, (x,t) + \ldots \tag{9}$$

where terms with subscript n are of $0 \, (\frac{a}{h})^{n-1}$. The time averaged values $\overline{q}$ and $\overline{\eta}$ are assumed to be of $0 \, (\frac{a}{h})$. For simple harmonic ocean tides, the first order solutions $q_1$ and $\eta_1$ are simple harmonic in time. Substituting the expressions for q and η, Eqs. (8) and (9), in Eqs. (6) and (7) yields, after time averaging and neglecting terms of $0 \, (\frac{a}{h})^2$ and higher

$$\overline{q} = \text{const} \tag{10}$$

$$F_\ell \overline{q} = - \frac{1}{h} \frac{\overline{\partial q_1^2}}{\partial x} - \frac{g}{2} \frac{\overline{\partial \eta_1^2}}{\partial x} - g \, h \frac{\partial \overline{\eta}}{\partial x} - F_\ell \frac{\overline{q_1 \eta_1}}{h^2} \tag{11}$$

The first two terms on the right represent the gradient of the tidally averaged momentum flux associated with the tidal velocities and the non-linear part of the pressure gradient, respectively. The third term is the gradient of the mean water

level, and the last term is a residual stress resulting from the non-linear bottom
friction. Together the terms on the right-hand side of Eq. (11) constitute the
producers of the residual flow $\bar{q}$.

For boundary conditions

$$\eta = a_o + a_1 \sin \omega t \qquad \text{at } x = -\frac{L}{2} \tag{12}$$

$$\eta = b_o + b_1 \sin (\omega t - \delta) \qquad \text{at } x = +\frac{L}{2} \tag{13}$$

where L = length of sea-level canal, it is shown in van de Kreeke and Dean (1975)
that the solution for $\bar{q}$ from Eqs. (10) and (11) is

$$\frac{\bar{q}T}{hL} = P \frac{a_1 b_1}{h^2} \sin \delta + Q \frac{a_1^2 - b_1^2}{h^2} + \frac{C_o h}{F_\ell L} \frac{\lambda}{L} \frac{a_o - b_o}{h} \tag{14}$$

P and Q are functions of the variables $\frac{C_o h}{F_\ell L}$ and $\frac{\lambda}{L}$. $C_o = \lambda/T$ is wave celerity for
zero friction, $\lambda$ = wave length for zero friction. $T = 2\pi/\omega$ = period of tidal wave.
The first term on the right-hand side of Eq. (14) represents the effect of the
phase difference, the second term the effect of the difference in tidal amplitude,
and the third term the effect of a difference in mean level on the residual flow.
The approximate analytical solution indicates among other results that

- the residual flow depends on that part of the boundary conditions which is
  difficult to measure, namely the mean water levels.

- in order for a residual flow to exist the mean water levels at the ends of
  the canal need not necessarily be different.

Using Eq. (14), values of the residual flow $\bar{q}$ were computed for various values
of the phase angle $\delta$. The length of the canal L = 20,000 m and the average depth
h = 3 m. Amplitudes of ocean tide are $a_1 = b_1 = 0.3$ m and mean levels $a_o = b_o = 0$.
The tidal amplitude T = 45,000 sec and the friction factor $F_\ell = 0.0025$ m/sec. The
results of the computations are presented in Fig. 1. In the same figure values of
the residual flow computed by numerically solving Eqs. (6) and (7) and time averaging
the discharge q are plotted.

The following is another way of illustrating the physics underlying the residual flow. Consider a single damped progressive wave in an infinite long canal,
Ippen (1966)

$$\eta_{1*} = a e^{-\mu x} \cos (\sigma t - kx) \tag{15}$$

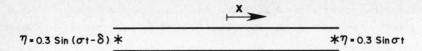

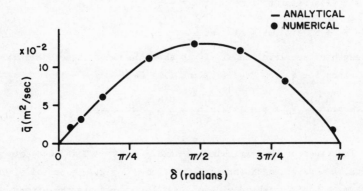

FIG. 1. RESIDUAL FLOW AS A FUNCTION OF
PHASE DIFFERENCE BETWEEN OCEAN
TIDES.

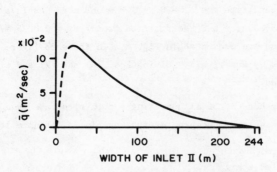

FIG. 2. RESIDUAL FLOW AS A FUNCTION OF
THE WIDTH OF INLET II.

$$q_{x*} = a \ c_o \ \frac{k_o}{\sqrt{\mu^2 + k^2}} \ e^{-\mu x} \ \cos (\sigma t - kx + \alpha) \qquad (16)$$

in which $k_o$ is the wave number for zero friction, and $\mu$, $k$ and $\alpha$ are defined by

$$k_o^2 = k^2 - \mu^2$$

$$\frac{F_\ell}{h} = (2\mu\sigma k)/(k^2 - \mu^2)$$

$$\alpha = \tan^{-1} (\frac{\mu}{k})$$

The subscript * refers to the infinite long canal. Substituting in Eq. (11), integrating with respect to x and setting $\bar{\eta}$ = o at x = o yields

$$\bar{\eta}_* = \frac{a^2}{h^2} \ \frac{k^2 h}{\mu^2 + k^2} \ (1 - e^{-2\mu x}) \qquad (17)$$

The mean water level in an _infinite_long_canal_ in the presence of a single damped progressive wave increases with increasing values of x. When dealing with a canal of _finite_length_ L, the value of the mean water level at x = L is forced to assume a certain value $\bar{\eta}_L$ prescribed by external conditions. This introduces a slope $\bar{\eta}_L/L$, which in turn gives rise to a uniform (= residual) flow. Details of this can be found in van de Kreeke (1971).

In addition to the tide characteristics at both ends of the canal, certain geometric features can play an important role in generating mass flows. For example, a sea-level canal freely connected to oceans having the same tide will not have a mean flow (because of the symmetry of the problem). When bay areas adjacent to the canal or when inlets with different dimensions at each end are added, the problem becomes asymmetric and the occurrence of a mean flow seems plausible. This is illustrated for the canal - inlet system in Fig. 2. The flow in the canal is described by Eqs. (6) and (7). The discharge in the inlets is assumed proportional to the cross-sectional area and the head difference across the inlet (linear inlet equation). As in the case of the sea-level canal, an analytical expression was derived for the canal - inlet system Cotter (1974), showing the independent effects of the inlets and tide characteristics on the mean current. As an example, values of analytically and numerically computed residual flows for various values of the width of Inlet II are presented in Fig. 2. In the example the amplitude of the ocean tides is 0.4 m, the tidal period is 45,000 sec, the average water depth is 3 m, the length and width of the canal are respectively 24,834 m and 762 m.

The length, width and depth of Inlet I are respectively 762 m, 244 m and 3 m. The length and depth of Inlet II are 762 m and 3 m. The friction factors are $F_\ell = 2.5$ $10^{-3}$ m/sec for the canal and $F_i = 2.10^{-3}$ m/sec for the inlets. Note that a residual flow exists, provided the width of Inlet I is different from the width of Inlet II. The foregoing example suggests that by properly selecting the dimensions of the inlets the residual flow can be considerably increased (van de Kreeke, 1976).

In the previous discussions, the description of the flow assumed linear friction and linear inlet equations. To obtain analytical solutions when including non-linear friction and non-linear inlet equations, friction and inlet equations need to be linearized. These linearization techniques are discussed in detail in Dronkers (1966). Applications to the problem of tide-induced residual flow can be found in van de Kreeke (1971) and van de Kreeke and Dean (1975).

## Residual Flows in Shallow Seas

Assuming a homogeneous incompressible fluid, a horizontal bottom, linear friction and neglecting the Coriolis force and lateral stress, the governing equations are

$$\frac{\partial \eta}{\partial t} + \frac{\partial q_x}{\partial x} + \frac{\partial q_y}{\partial y} = 0 \tag{18}$$

$$\frac{\partial q_x}{\partial t} + \frac{2q_x}{h} \frac{\partial q_x}{\partial x} + \frac{1}{h} \frac{\partial q_x q_y}{\partial y} = \frac{-F_\ell q_x}{h + \eta} \tag{19}$$

$$\frac{\partial q_y}{\partial t} + \frac{2}{h} q_y \frac{\partial q_y}{\partial y} + \frac{1}{h} \frac{\partial q_x q_y}{\partial x} = \frac{-F_\ell q_y}{h + \eta} \tag{20}$$

The number and type of boundary conditions necessary to formulate a well-posed problem is still very much a matter of debate Dronkers (1975), Gerritsen (1977). Here the boundary conditions commonly found in the engineering literature will be used i.e.

at closed boundaries:   normal velocity is zero

at open boundaries:   water levels are prescribed

Equations (18), (19) and (20) are solved numerically using an explicit finite difference scheme. A typical space-step for the model is $\Delta s = 800$ m and the time step $\Delta t = 75$ sec. Once the instantaneous discharges $q_x$ and $q_y$ are known the residual flow is found by averaging over the tidal cycle.

In the finite difference formulation of Eqs. (18) - (20), space and time derivatives are approximated by central difference quotients. At the closed boundaries, the derivative part of the cross-differentiated terms in the convective acceleration is approximated by a forward difference (this is equivalent to taking the second derivative equal to zero). To accomodate the cross-differentiated terms at the open boundaries three often used approaches are considered

Case a. velocities parallel to the boundary are taken equal to zero; in that case the cross-differentiated terms at the boundary do not enter the computation.

Case b. the derivative part of the cross-differentiated terms is approximated by a backward difference.

Case c. the derivative part of the cross-differentiated term is taken equal to zero.

The approximations made in accomodating the cross-differentiated terms at the open boundaries could be construed as additional boundary conditions. They will be further referred to as secondary boundary conditions.

To test the numerical model a semi-analytical solution for the residual flow was developed for a square bight with a horizontal bottom. The perturbation technique discussed in the previous section was applied but for two-dimensional flow fields led to unwieldy expressions. Instead, it was decided to start from a relatively simple first order solution consisting of a combination of damped progressive waves travelling in the x and y direction and to compute the corresponding boundary conditions. The assumed first order solution for $\eta$, $q_x$ and $q_y$ and the derived boundary conditions are presented in Appendix A. To find the residual flow, the non-linear terms in Eqs. (19) and (20) are approximated by substituting the first order expressions for the water level and the tidal flows. Averaging over the tidal period and eliminating the residual flows in the x and y direction between the Eqs. (18), (19) and (20) yields a Poisson-type equation for the residual water level. Assuming the residual water level to be equal to zero at the open boundary, the Poisson equation is solved using a standard finite difference technique. Once the residual water levels are known the residual flows in the x and y directions follow by numerically differentiating the residual water levels with respect to x and y. An example of residual flow patterns using this method is presented in Fig. 3A.

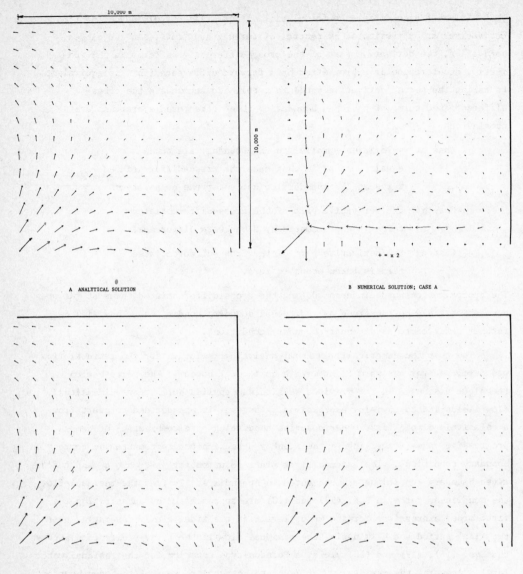

A  ANALYTICAL SOLUTION

B  NUMERICAL SOLUTION; CASE A

C  NUMERICAL SOLUTION; CASE B

D  NUMERICAL SOLUTION; CASE C

LEGEND

FLOW SCALE $\longmapsto$ 80 cm$^2$/sec

a = 0.2 m        T = 45,000 sec

h =   3 m

FIG. 3.  RESIDUAL FLOW IN
SQUARE BIGHT

Using the same boundary conditions the residual flow pattern for the square bight is computed numerically. Computations are carried out for the three different secondary boundary conditions. The results are presented in Figs. 3B, 3C and 3D. The numerically computed residual flow for cases b and c compare well with the semi-analytically computed residual flow. The same holds for case a when restricting attention to the interior (starting at approximately two grid points or 1600 m from the open boundary). Close to the open boundary the numerically computed residual flow differs considerably from the true (= semi-analytical) solution. An explanation for this is that the secondary boundary condition for case a is inconsistent with the true solution. The velocities parallel to the boundary are not zero as follows from Appendix A, Eq. (7). For case b and c, the secondary boundary conditions are also inconsistent with the true solution. However, in these cases the inconsistency is of a second order nature; the first order expressions for

$$\frac{\partial q_y}{\partial y}, \frac{\partial q_y}{\partial x}, \frac{\partial^2 q_x}{\partial y^2} \text{ and } \frac{\partial^2 q_y}{\partial x^2} \text{ are zero everywhere.}$$

One of the more obvious differences between the numerical solution with case a as secondary boundary condition and the true solution is the occurrence of closed flow lines (eddies) in the residual flow along the open boundaries. Assuming that the flow lines closely resemble the path lines,(or equivalently the second term on the righthand side of Eqs. (4a) and (4b) is small compared to the first term) this implies that the water in the interior does not exchange with the ocean water; an important practical consequence, which demonstrates the importance of the secondary boundary conditions when computing residual flows.

Acknowledgement

This study was supported by the National Science Foundation under Grant No. ENG-76-08288.

Appendix A

## ANALYTICAL SOLUTION FOR THE TIDAL FLOW IN A RECTANGULAR BIGHT

### Geometry; Coordinate System

The geometry of the bight and the cocordinate system used in the mathematical formulation of the flow are described in Fig. 1. Assumed is a horizontal bottom.

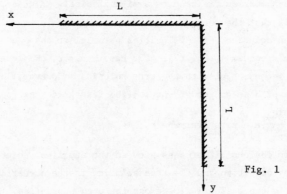

Fig. 1

### Equations

In the governing equations the non-linear terms are neglected and the friction term is linearized.

$$\frac{\partial q_x}{\partial t} + gh \frac{\partial \eta}{\partial x} = - \frac{F_\ell q_x}{h} \tag{1}$$

$$\frac{\partial q_y}{\partial t} + gh \frac{\partial \eta}{\partial y} = - \frac{F_\ell q_y}{h} \tag{2}$$

$$\frac{\partial \eta}{\partial t} + \frac{\partial q_x}{\partial x} + \frac{\partial q_y}{\partial y} = 0 \tag{3}$$

in which

$q_x$ = flux per unit width in x direction

$q_y$ = flux per unit width in y direction

$\eta$ = water level with respect to Still Water Level

$h$ = depth with respect to Still Water Level

$t$ = time

$F_\ell$ = friction coefficient for linearized friction

## Boundary Conditions

The water level at the open boundaries are selected such that the resulting solution consists of the sum of two progressive waves propagating in the positive and negative x-direction and two progressive waves propagating in the positive and negative y-direction. This yields for

$$x = L$$

$$\eta = a \cos \sigma t + \frac{2a}{\cosh 2\mu L + \cos 2kL} \Big[ (\cosh \mu y \cos ky \cosh \mu L \cos kL$$

$$+ \sinh \mu y \sin ky \sinh \mu L \sin kL) \cos \sigma t - (\sinh \mu y \sin ky \cosh \mu L$$

$$\cos kL - \cosh \mu y \cos ky \sinh \mu L \sin kL) \sin \sigma t \Big] \tag{4}$$

For

y = L, $\eta$ is the same as Eq. (4) with y replaced by x. In Eq. (4), 2a is the amplitude of the tide at (x = L, y = L). $\sigma$ is the angular frequency of the tide.

$$k^2 - \mu^2 = k_o^2 \qquad k_o = \sigma/\sqrt{gh} \qquad F_\ell/h = 2\mu\sigma k/(k^2 - \mu^2)$$

At the closed boundaries the flux perpendicular to the boundary is zero.

$$q_x = 0 \quad \text{for } x = 0$$

$$q_y = 0 \quad \text{for } y = 0 \tag{5}$$

## Solution

$$\eta = \frac{2a}{\cosh 2\mu L + \cos 2kL} \Big\{ \Big[ \cosh \mu L \cos kL \ (\cosh \mu x \cos kx + \cosh \mu y \cos ky)$$

$$+ \sinh \mu L \sin kL \ (\sinh \mu x \sin kx + \sinh \mu y \sin ky) \Big] \cos \sigma t$$

$$- \Big[ \cosh \mu L \cos kL \ (\sinh \mu x \sin kx + \sinh \mu y \sin ky) - \sinh \mu L \sin kL$$

$$(\cosh \mu x \cos kx + \cosh \mu y \cos ky) \Big] \sin \sigma t \Big\} \tag{6}$$

$$q_x = \frac{2a \ C_o \ \Phi}{\cosh 2\mu L + \cos 2kL} \Big[ - (\sinh \mu x \cos kx \cosh \mu L \cos kL + \cosh \mu x$$

$$\sin kx \sinh \mu L \sin kL) \cos (\sigma t + \theta) + (\cosh \mu x \sin kx \ \cosh \mu L$$

$$\cos kL - \sinh \mu x \cos kx \sinh \mu L \sin kL) \sin (\sigma t + \theta) \Big] \tag{7}$$

The expression for $q_y$ is the same as Eq. (7) with x replaced by y. In Eq. (7):

$$\Phi = \frac{k_o}{\sqrt{\mu^2 + k^2}} \qquad\qquad \theta = \tan^{-1} (\tfrac{\mu}{k}) \qquad\qquad C_o = \sqrt{gh}$$

References

Cotter, D. C. "Tide-Induced Net Discharge in Lagoon-Inlet Systems". Technical Report, UM-RSMAS #74031. University of Miami, Coral Gables, Fla., p. 39, 1974.

Dronkers, J. J. "Tidal Computations in Rivers and Coastal Waters". North-Holland Publishing Co., Amsterdam, p. 518, 1966.

Gerritsen, H. "Solution Methods for the Shallow-Water Equations I". Memorandum No. 1968. Department of Applied Mathematics, Twente University of Technology, Enschede, Netherlands, p. 31, 1977.

Ippen A. "Estuary and Coastline Hydrodynamics". McGraw-Hill Book Company, Inc. p. 744, 1966.

Longuet-Higgins, M. S. "On the Transport of Mass by Time-Varying Ocean Currents". Deep Sea Res., Vol. 16, pp. 431-447, 1969.

Nihoul, J. C. F. and F. C. Ronday. "The Influence of the Tidal Stress on the Residual Circulation". Tellus XXVII, pp. 284-489, 1975.

Tee, K. T. "Tide-Induced Residual Current, a 2-D Non-linear Numerical Tidal Model". Journal of Marine Research 34, 4, pp. 603-628, 1976.

van de Kreeke, J. "Increasing the Mean Current in Coastal Channels". J. of the Waterways, Harbors and Coastal Engineering Division, ASCE, Vol. 102, No. WW2, pp. 223-234, 1976.

van de Kreeke, J. and R. G. Dean. "Tide-Induced Mass Transport in Lagoons". J. of the Waterways, Harbors and Coastal Engineering Division, ASCE, Vol. 101, No. WW4, pp. 393-403, 1975.

van de Kreeke, J. "Tide-Induced Mass Transport in Shallow Lagoons". Technical Report No. 8, Dept. of Coastal and Oceanographic Engineering, University of Florida, Gainesville, Fla. p. 111, 1971.

SIMULATION OF TIDAL RIVER DYNAMICS

K.-P. Holz

Chair of Fluid Mechanics
Technical University Hannover
Federal Republic of Germany

## Summary

The catalogue of engineering problems encountered in tidal rivers is
quite manifold and demands different modelling techniques. All of them
have their particular ranges of application, and do not give reliable
results if the assumptions made are too crude. In which way limitations
come in will be shown for a storm surge simulation, for which one- and
two-dimensional numerical models as well as a combined hydraulic-numer-
ical model were applied. The last one proves to be particularly suited
for the treatment of local effects.

## Introduction

The prediction of changes, caused by human interference in the natural
regime of a tidal river, is of great practical interest for preserving
or improving conditions of navigation, industrialization, environment,
and for storm surge protection. The latter is of particular interest
in the biggest German tidal river Elbe, which gives access to the port
of Hamburg. So extensive modelling is being made on this river, which
gives the possibility of critically surveying some of the used model-
types.

## Formulation of the Problem

The Elbe river connects the North Sea with the port of Hamburg (Fig. 1).
The distance from the open sea at Cuxhaven to the harbour is about 11o
kilometers. In this part of the river extensive dredging was made in
the past. Due to growing ship units, the navigation channel was deepened
to 13.5 m. At the same time, land reclamation was made and new dikes
were built in many areas. As a consequence, the cross-section of the
river under extreme high-water conditions was reduced. Moreover, upstream Hamburg harbour a weir was built in order to control the navig-

ation conditions in the upstream waterway. This construction led to a
further reduction of the tidal volume of the river, as it cuts off a
part of 9o km length of the river between the weir at Geesthacht and
Neu-Darchau, which was originally tide-influenced. Both modifications
led to higher and faster-running storm surges on the river.

Since in February 1962 a great area in the Hamburg harbour was flooded,
engineers and politicians have been discussing means to protect it from
being flooded again. Among many other aspects, the questions mainly concerned an optimal control of the upstream weir, and a controlled opening of dikes in order to cut off the peak of a storm surge in the Hamburg harbour area. The models applied in connection with these problems
will be briefly presented and discussed.

The One-dimensional Model

To get a first insight into the behaviour of the river under storm-surge
conditions, a one-dimensional numerical model was set up. It was felt,
that this model, which runs at reasonably low costs, was the first step
and at the same time the best tool to obtain information on whether
more sophisticated models were needed.

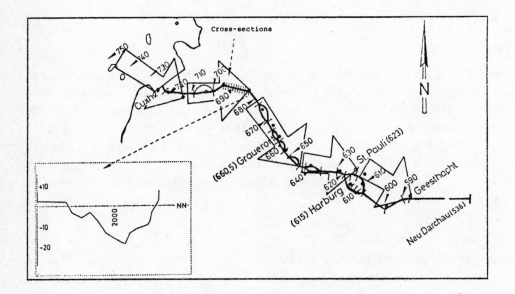

Fig. 1: One-dimensional Model of the Elbe River

The one-dimensional approach must be understood as a tool for a far-
field simulation which does not include local effects over the cross-
section of the river. The model covers a river length of about 19o km.

The cross-sections between the dikes have a width of about 5oo m in the upstream part of the river. They widen up to about 15.ooo m at the river mouth. The river downstream the Hamburg harbour area is characterized by banks and islands, so that a branched model had to be used.

The describing equations for the numerical model were derived from the conservation principles of physics.

Momentum conservation:

$$\frac{D}{Dt}\int_V v_i \rho \, dv - \int_V f_i \rho \, dv - \int_A \tau_{ij} n_j \, dA = 0 \tag{1}$$

    inertia     gravity     bottom-, surface stress

Mass conservation:

$$\frac{D}{Dt}\int_V \rho \, dv = 0 \tag{2}$$

Assuming an integrated velocity v over the cross-section, and a mean flux Q, the well-known equations for open-channel flow are obtained[1].

$$\int \left( \frac{1}{gA} \frac{\partial Q}{\partial t} + \frac{1}{g} \frac{Q}{A^2} (\alpha + \frac{b_s}{b}) \frac{\partial Q}{\partial s} + (1 - \frac{\alpha}{g} \frac{Q^2}{A^3} b_s) \frac{\partial h}{\partial s} - I_s + I_v \right) ds = 0 \tag{3}$$

$$\int (b\frac{\partial h}{\partial t} + \frac{\partial Q}{\partial s} - \bar{q}) \, ds = 0 \tag{4}$$

They are given in an integrated formulation. The notation used is shown in Fig. 2.

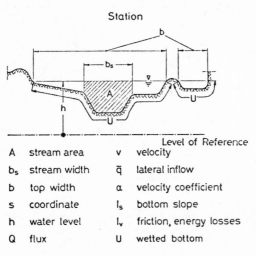

| | | | |
|---|---|---|---|
| A | stream area | v | velocity |
| $b_s$ | stream width | $\bar{q}$ | lateral inflow |
| b | top width | $\alpha$ | velocity coefficient |
| s | coordinate | $I_s$ | bottom slope |
| h | water level | $I_v$ | friction, energy losses |
| Q | flux | U | wetted bottom |

Fig. 2: Cross-Section and Notation

The formula of Manning was used for the friction. k is the friction parameter, D stands for the hydraulic radius D = 4 U/A.

$$I_R = \frac{|v|v}{k^2 (\frac{D}{4})^{4/3}} \tag{5}$$

For the simulation of storm surges, the wind influence has also to be included in the model. A formulation analogous to the Taylor bottom friction term was set up. $\mu$ is a parameter and w stands for the wind velocity component in the flow direction of the river.

$$I_w = - \frac{\mu|w|w}{gh} \tag{6}$$

The numerical solution of equ. (3,4) can be performed either by the finite difference or the finite element method. If the integral formulation (3,4) is reduced, any standard finite difference technique can be used for the remaining differential form. The presented model applies the discretization proposed by PREISSMANN [2]. The numerical properties of this formulation were carefully tested by EVANS [3].

finite difference method
[D] = 0

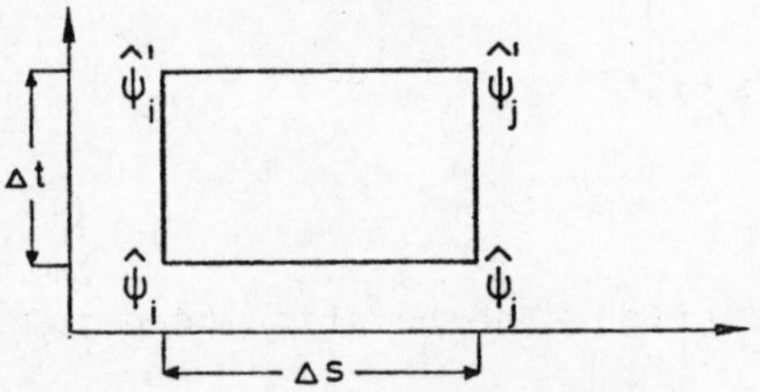

$$\frac{\partial \psi}{\partial t} = \frac{1}{2\Delta t} (\hat{\psi}'_i - \hat{\psi}_i + \hat{\psi}'_j - \hat{\psi}_j)$$

$$\frac{\partial \psi}{\partial x} = \frac{\theta}{\Delta x} (\hat{\psi}'_i - \hat{\psi}'_j) + \frac{(1-\theta)}{\Delta x} (\hat{\psi}_i - \hat{\psi}_j)$$

$$\psi = \frac{1}{2} (\theta (\hat{\psi}'_i + \hat{\psi}'_j) + (1-\theta)(\hat{\psi}_i + \hat{\psi}_j))$$

Fig. 3: Discretization by the Finite Difference Method

On the other hand, one can start from the conservation formulation of the equ. (3,4) directly. This will then lead to the finite element approach. Incidentally, it can be seen that no special technique, as e.g. the Galerkin method, is required for the solution. The direct integration of the equations

$$\int [D] \, ds = 0 \tag{7}$$

with the trial functions

$$\bar{\psi}(s) = \sum_i \theta_i(s) \, \hat{\psi}_i \tag{8}$$

for a state variable $\psi$ leads to the following equation in terms of the method of weighted residuals

$$\int ( [ D(\psi(s)) ] - [D(\bar{\psi}(s)) ] ) \, ds = 0 \qquad (9)$$

From this follows

$$\int [ D(\bar{\psi}(s)) ] \, ds = 0 \qquad (1o)$$

which can also be interpreted as the subdomain approach with a constant weighting function equal to one.

The finite difference as well as the finite element formulations lead to an implicit equation system. For its solution, a special double sweep technique was applied. For any branch of the system, two equations are formulated, in each of which the flux at one end-point depends on the water-level quantities alone. The mass conservation law, applied at each branching point, leads to an equation system which has the water level as the only unknown quantity. Thus the overall equation system continues to be rather small. The algorithm can be run on small computers and is quite fast.

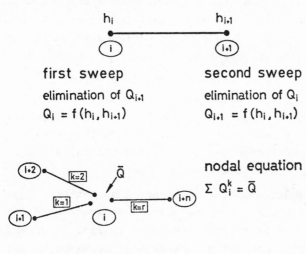

**first sweep**
elimination of $Q_{i+1}$
$Q_i = f(h_i, h_{i+1})$

**second sweep**
elimination of $Q_i$
$Q_{i+1} = f(h_i, h_{i+1})$

**nodal equation**
$\sum Q_i^k = \bar{Q}$

Fig. 4: Solution Procedure and
Equation System

## Application of the One-dimensional Model

The described model was applied with a discretization as given in Fig. 1. Cross-sections were taken at each kilometer. The time step used for calculation was 1o minutes, which corresponds to a Courant number of 12 to 18.

Some specific approximation had to be made when taking up the cross-section profiles (Fig. 5). In some areas banks and islands exist under normal tidal conditions, ranging from -1.1o m to 1.6o m at the station shown in Fig. 5. Under these conditions, the system has to be divided

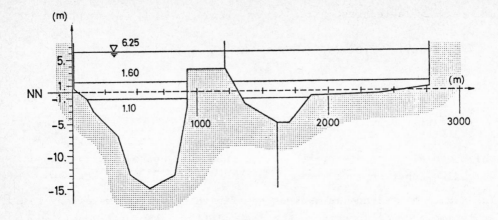

Fig. 5: Cross-Section km 642.o, Elbe River

into two branches. For extremely low water, the shallow branch may even
dry out during ebb flow. This condition can easily be simulated by providing the cross-section with a narrow slit, in which a water level can
always be computed, while the computed flux is negligeably small. No
additional software has to be implemented.

On the other hand, the defined two branches should be united to form
one branch for extreme high-water situations. This would require additional software complications, since the discretization should change
dynamically then. To avoid this, a wall was assumed on the bank. A
comparison between the water levels in both cross-sections, which
practically did not differ, justified this assumption.

The model was calibrated under normal tidal conditions. It became evident that the friction parameter had to be made flow-direction-depen-
dent. The values of this parameter which are given in Fig. 6 correspond
to the natural morphological conditions. Downstream the Hamburg harbour
area, typical alluvial conditions are found. In the harbour area itself,
a dramatic increase of friction is observed. This is due to a very crude
approximation of the highly complicated system of harbour basins by a
few branches only. The area upstream Hamburg is characterized by more
regular profiles and more uniform flow conditions. The agreement of the
friction distribution with the natural conditions shows that the discretization was chosen at a reasonable scale.

Moreover, it should be mentioned that the total cross-sections were
used for the computation. A reduction to hydraulically effective cross-
sections was not necessary. This raised good hope for satisfactory
simulations of storm surges. The accuracy obtained from the calibration

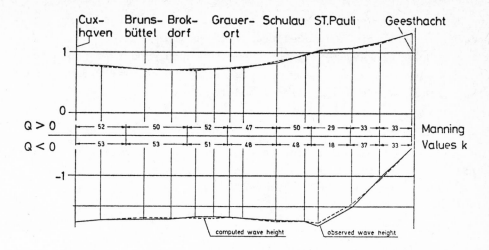

Fig. 6: Friction and Tidal Range Calibration of Elbe River

is shown in Fig. 6. The maximum error at high- and low-tide was less than 7 cm and the phase shift was about 5 minutes.

The calibrated model was then applied to storm surge simulations. To get a reasonable representation for the time- and space-dependent wind-field, the area of the Elbe was split up into 5 regions (Fig. 1), for which different wind data from stations in these regions were specified. Fig. 7 shows the water level history at St. Pauli, a place in the Hamburg harbour area. The simulation was run with the observed water-level variation at Cuxhaven as a prescribed boundary condition, and a constant inflow at the upstream boundary. The agreement between observed and computed data is excellent.

This encouraging result leads to further simulation runs for different purposes. Studies were made with regard to an optimal control at the weir upstream Hamburg, as well as the influence of a protection dam downstream the harbour, which can be closed in the event of storm surges.

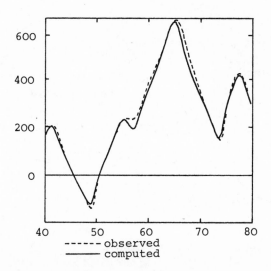

Fig. 7: Storm Surge 1.-4. Jan.1976
Water Level Variation at St. Pauli

The question whether a removal or a controlled opening of dikes could reduce the observed high water peaks was of great interest. This question could not be answered by the one-dimensional model as the process of flooding the land behind the dikes cannot be parametrized and must be simulated by a two-dimensional model at least.

## The Two-dimensional Model

The two-dimensional model uses an explicit finite difference formulation [4]. It starts from the well-known differential equations for vertically integrated flow [1].

$$\frac{\partial u}{\partial t} + u\frac{\partial u}{\partial x} + v\frac{\partial u}{\partial y} - fv = -g\frac{\partial h}{\partial x} + A_H \Delta u + \frac{\tau_{ob}^{(x)} - \tau_{Bd}^{(x)}}{a + h} \tag{11}$$

$$\frac{\partial v}{\partial t} + u\frac{\partial v}{\partial x} + v\frac{\partial v}{\partial y} + fu = -g\frac{\partial h}{\partial y} + A_H \Delta v + \frac{\tau_{ob}^{(x)} - \tau_{Bd}^{(x)}}{a + h} \tag{12}$$

$$\frac{\partial h}{\partial t} + \frac{\partial}{\partial x}((a+h)u) + \frac{\partial}{\partial y}((a+h)v) = 0 \tag{13}$$

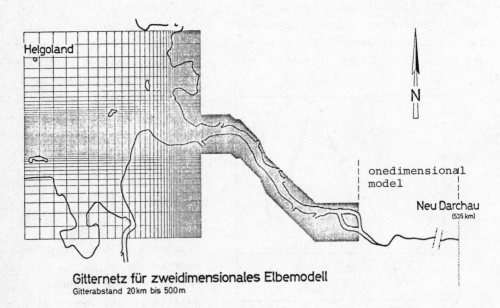

**Gitternetz für zweidimensionales Elbemodell**
Gitterabstand 20 km bis 500 m

Fig. 8: Discretization Two-dimensional Model, Elbe River

The model was extended to the inner part of the German Bight. Thus it was possible to study the interaction between changes in the river dynamics and the flow conditions on the intertidal flats in the bight, which was not possible in the one-dimensional formulation.

The discretization in the river itself had a resolution of 5oo m up to the harbour area. The upper end of the river was simulated by a one-dimensional model again. Boundary conditions were prescribed at the open sea. The calibration of this model turned out to be equally or even more complicated compared to that for the one-dimensional model. The simulation of the already mentioned storm surge of January 1. - 4., 1972, finally gave results as satisfactory as the one-dimensional model did (Fig. 7).

The two-dimensional model was then applied to the dike-breaking and controlled dike opening problem. Though it was now possible to simulate the flooding process behind the dike more realistically, it was found that the discretization was too crude. A refinement of the grid, however, would further increase the computing costs, which were already about 3o times higher than those for the one-dimensional simulation. A realistic alternative could be given by a one-dimensional system, to which two-dimensional subsystems were fitted for the open sea area and for the dike investigation areas. This approach, however, would only reduce the computing costs of the model, but not help to overcome a more important difficulty. This results from the fact that for disastrous situations, as in the event of dike-breaking, practically no measurements or observations are available which could help to calibrate a model. Therefore, in this situation, the value of numerical simulations is rather limited, and it is felt that more information could be obtained from hydraulic models.

The Hybrid Model

It cannot be expected that hydraulic models are an alternative to numerical ones under all conditions, since they too have their limitations. This is not only due to the costs and laboratory space needed but also to physical restrictions which come in from the model laws. Especially in a case as the one discussed here, where the entire river has to be modelled, a very small scale has to be chosen. It then becomes doubtful whether the friction is still represented correctly. It can be hoped, however, that the answers of a hydraulic model simulation may become more reliable when the local situation of the land behind the dikes can be modelled with a high resolution at a reasonable scale.

The main problem is to define the scale and to find out how big and costly such a model would become. A scale of 1:3oo for the horizontal dimension, with a distortion factor of 3 for the vertical dimension, was judged to be sufficient for the intended application. This scale would result in a model of about 63o meters for 19o km of river length, an unrealistically big size. As the main intention was to simulate the dike problem in an area between Altenbruch, at km 72o near Cuxhaven, and Grauerort (Fig. 1), only the upstream river part had to be modelled for giving the correct system behaviour. So a numerical model could also be used for this purpose, if it dynamically interacted with the hydraulic model.

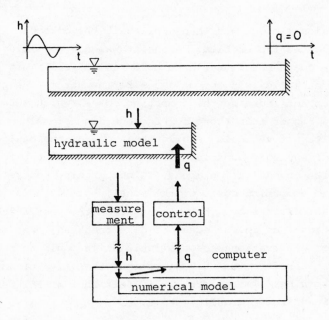

Fig. 9: Principle of Hybrid Model

The basic idea is shown in Figure 9. The hydraulic model of the entire system is cut into two parts, a hydraulic one and a mathematical one. Both models are run under real-time-conditions. At the connection point, the water level h is measured in the hydraulic model, and then given as boundary condition for the numerical model. By this the discharge is now computed and then controlled on the hydraulic model. Measurement and control are repeated in small time-steps, so that practically continuous conditions are obtained at the coupling point [5,6]. A model of this type is called a hybrid model.

The described technique was applied to the Elbe river. The hydraulic part covered the area between Altenbruch and Grauerort and had a length of 2oo m. The tide, controlled at the boundary at Altenbruch, had a period of 24 minutes. The coupling interval was 1o seconds. Its value was determined from the minimum computing time for the numerical model, for which the described and already calibrated one-dimensional formulation was used. Figure 1o shows the error for high and low water,

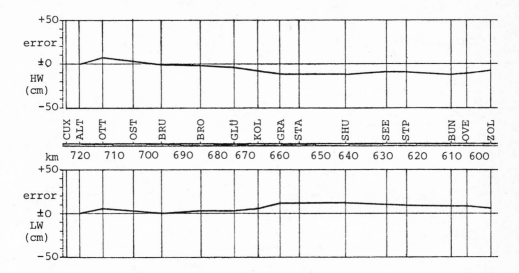

Fig. 1o: Hybrid Model Elbe River

when the one-dimensional numerical and the hybrid models are compared for a normal tide. The errors are beyond 1o cm, which, in comparison to the total tidal range at the corresponding places, is of about 7%. This magnitude was accepted; it could possibly be reduced further by a more careful calibration in the hydraulic part.

The hybrid model was then applied to practical problems. It became obvious that the areas which originally were intended to be flooded, were too small and could not help to cut down significantly the storm-surge peak. Other investigations are still going on to find out whether dams or cross-section reductions might have a stronger influence. In all these cases, the high resolution of topography which is given by the hydraulic model, and the three-dimensional simulation of flow conditions, proved to be advantageous, especially with respect to the analysis of near field conditions.

Conclusion

The storm-surge simulation on the river Elbe was chosen to demonstrate the range of applicability of one- and two-dimensional numerical models. For investigations, when practically no calibration of the models can be made, and near-field conditions prevail, these models give limited answers only. The models can be improved then by combining them with hydraulic model parts, which represent those areas, for which more detailed information is wanted. The combination of numerical and hydraulic models, which are coupled real-time, reduces the size of a purely hydraulic model, as well as the costs, and leads to three-dimensional simulations.

<u>References</u>

1. Dronkers, J. J., "Tidal Computations in Rivers and Coastal Waters", North Holland, Amsterdam, 1964

2. Preissmann, A., "Propagation des intumenscences dans les canaux et les rivières", $1^{er}$ Congrès de l'Association Française de Calcul, Grenoble 1960

3. Evans, E. P., "The Bahaviour of a Mathematical Model of Open Channel Flow", Proc. 17th IAHR-Congres, Vol. 2, pp. 173 - 180, 1977

4. Maier-Reimer, E., Backhaus, J., "A Combined Model for the German Bight and River Elbe" (in preparation)

5. Holz, K.-P., "Analysis of Time Conditions for Hybrid Tidal Models", ASCE, Proc. 15th Conf. Coastal Engineering, Vol. IV, pp. 3460 - 3470, 1976

6. Funke, E. R., Crookshank, N. L., "A Hybrid Model of the St. Lawrence River Estuary", ASCE, Proc. 17th Conf. Coastal Engineering

ANALYSIS OF TIDE AND CURRENT METER

DATA FOR MODEL VERIFICATION

John D. Wang
Ocean Engineering
Rosenstiel School of Marine & Atmospheric Science
4600 Rickenbacker Causeway
Miami, Florida 33149

## Introduction

Mathematical models are increasingly being used to describe the evolution of hydrodynamic field and/or water quality parameters in coastal waters. Irrespective of the specific features of a model, whether it be a hurricane surge model for an open coast, a model of continental shelf circulation, or a model for tidal flow in a shallow coastal embayment, it is reasonable to assume that its ultimate use will be closely connected with the design of certain human-related activities. These include exploration of natural resources, waste disposal, shore protection etc., and usually are of significant economic importance. To make sensible planning and design decisions it is therefore essential that the accuracy and sensitivity of calculated model results be well understood.

Due to user inexperience, simplifying assumptions, idealisation of topography and treatment of boundary conditions in models it is possible to introduce inaccuracies in computed results. In order to establish confidence in a model it is therefore necessary to verify such results against real data.

In this paper we outline some data analysis techniques for verifying different aspects of a two-dimensional finite element hydrodynamic model that solves the vertically integrated equations of motion. This model is described by Wang (1978).

The use of time series analysis for model calibration is introduced by Leendertse and Liu (1974), and these techniques are being extended here to make the comparison between model predictions and field observations more meaningful. In most studies done sofar the emphasis has been more on numerical aspects, while verification is commonly limited to a qualitative (often subjective) comparison of surface elevations and, on rare occasions, also of velocities. By using the proposed techniques it will be possible to make the verification process more quantitative and hopefully objective. It will also become clear that the usefulness of the approach is not limited to the specific model chosen for illustration here.

## What to Verify

Since the governing equations on which the model is based are derived exactly from fundamental physical laws the verification process can be quickly reduced to investigating the accuracy and validity of simplifying assumptions and empirical parameterizations. These are defined as any relationship that is not derived exactly from a fundamental physical law. To simplify the following discussion we shall neglect internal stresses and any errors that could be introduced by the numerical technique.

In the present situation, i.e. for a model based on the vertically integrated equations of motion, the phenomena that need to be verified are

1. Vertical uniformity of velocities
2. Bottom friction parameterization
3. Wind stress parameterization
4. Boundary conditions.

The hydrostatic pressure assumption and assumption of incompressibility could be added to the list, however their validity is generally accepted. In the following sections suggestions are made for verification procedures.

Vertical Velocity Distribution

The existence of a fairly uniform vertical velocity distribution is a basic assumption required for predictive modeling (non-uniform velocity distributions could in theory be handled by calibration of eddy viscosities, but the model would not be predictive), and should be validated in the field before any modeling attempt. For shallow, unstratified water bodies, experience has shown that tidal flow most often satisfies this condition, however when wind is superposed the situation may be drastically changed. In general it is necessary to quantify the deviation from vertical homogeneity in order to assess the possible effect on the accuracy of model results.

Non-uniformity may take the form of significant vertical shear outside the near bottom region, or rotation of the velocity vector about a vertical axis.

In order to model wind driven flow accurately the time scale of vertical momentum diffusion must be small such that the surface boundary layer quickly reaches the bottom. However, because of the possibility of establishing counterflow situations in confined water bodies, this is not a sufficient condition and additional verification is necessary.

Vertical velocity profiles should be measured to assure negligible shear, but in addition long term current records are needed to enable differentiation between wind driven and tidally driven flow. The occurrence of short period surface waves during wind events make current measurements difficult in shallow coastal waters. Ideally a fast responding instrument such as an electromagnetic current meter should be used to be able to discern the wave motion in the velocity record. Unfortunately, the wave field is generally three-dimensional and random. The measurement of such a wave field is at the margin of present technology and beyond the scope of this exposition.

In a verification study for Biscayne Bay, Florida, Fig. 1,

159

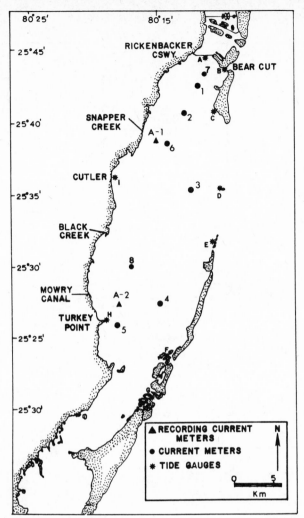

FIGURE 1 — BISCAYNE BAY. LOCATION OF STATIONS

a number of conventional anemometer type, Teledyne-Gurley, current meters were used to measure current profiles, Fig. 2.

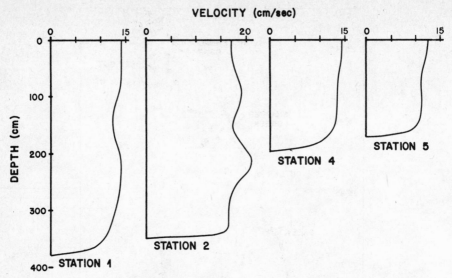

Fig. 2 - Velocity profiles in Biscayne Bay

In addition a recording savonius rotor, Aanderaa, current meter was deployed at locations in the bay for periods of 4 and 3 weeks.

The Teledyne-Gurley current meters could not be used reliably during windy weather and only one Aanderaa meter was available making it impossible to obtain vertical profiles under windy conditions. The Aanderaa meter was placed four feet below the surface and recorded currents during different wind and surface wave conditions. However, from the Aanderaa records, Fig. 3,

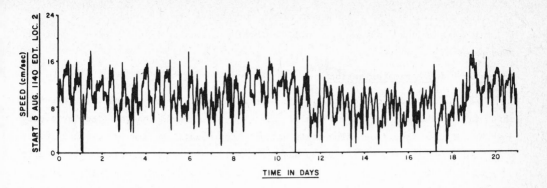

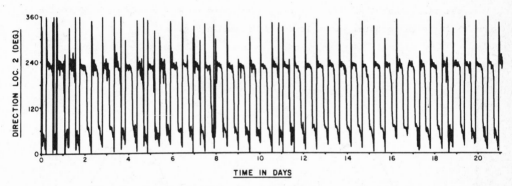

Fig. 3 - Speed and Direction Data for Current Meter A-2

it appears that wind driven velocity components are small compared to the tidal currents, at least when judging from the consistency of the direction record.

For the vertical velocity profiles a momentum coefficient was determined from

$$\alpha = \frac{\int_{-h}^{\eta} U^2 dz}{(h+\eta)V^2} \tag{1}$$

where U is the measured speed and V is the vertically average velocity determined by

$$V = \frac{1}{h+\eta}\sqrt{\left(\int_{-h}^{\eta} U\cos\theta dz\right)^2 + \left(\int_{-h}^{\eta} U\sin\theta dz\right)^2} \tag{2}$$

and $\theta$ is the direction of U.

Thus $\alpha$ provides an estimate of the deviation of the measured velocity profile from a plane vertically uniform profile.

For Biscayne Bay a range of $\alpha$ from 1.1 to 1.25 was found with larger values typically occurring near slack tides. This compares with a value of 1.08 assuming a plane logarithmic profile for rough flow with depth = 50 x roughness.

The error in assuming a vertically homogeneous velocity profile appear to be small in this case.  The effect of the error on the accuracy of computed flow fields under windy conditions could not be determined, since it would require more accurate field measurements.

## Bottom Friction Parameterization

The treatment of bottom friction in numerical models rely on empirical relationships such as Manning or Chezy formula.  Although these have been verified for steady uniform and uni-directional flow their validity for coastal waters is not automatic.

Traditionally, the verification and calibration of bottom friction in vertically integrated models have been carried out using surface elevations, since these are easily measured.  Bottom friction influences both tidal range and phase lags so that it is possible to adjust the friction coefficients to make surface elevations agree. Experience has unfortunately shown that surface elevation changes are relatively insensitive to friction while velocities are more sensitive.  Thus it is not unusual that predicted tidal ranges and phase lags are accurate to within 10% with minimal adjustment of friction factors.  This is demonstrated in Fig. 4

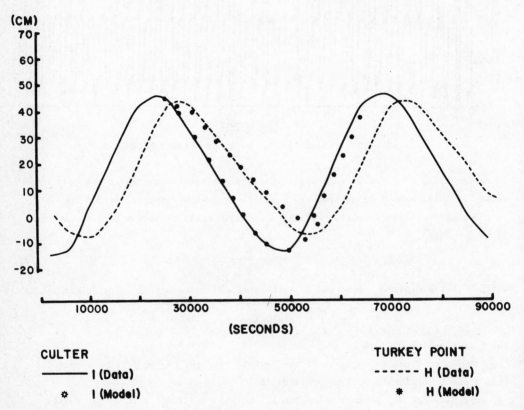

Fig. 4 - Comparison of Surface Elevations

which shows a comparison of measured and predicted tidal elevations at Cutler and

Turkey Point in Biscayne Bay, (see Fig. 1). Manning's equation was used for prediction with Mannings n = .025, a commonly used value, and no attempts were made to adjust n. Tide ranges at the two stations were accurate to within 9% and phase lags to within 2%. Using an n value of .03 these figures changed to 8% and 4% respectively.

Verification of surface elevations is a convenient preliminary check of model performance, however for proper verification a comparison of velocities is needed.

An example of computed and measured hodographs is shown in Fig. 5

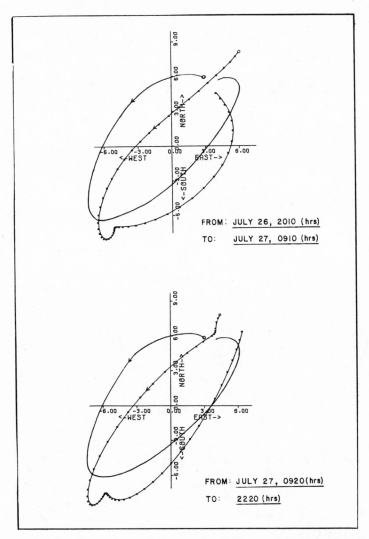

Fig. 5 - Comparison of Computed and Measured Hodographs

and the difficulty in making a simple and meaningful quantitative comparison becomes apparent. In general the criteria to be used must be determined from the ultimate objective of the model, e.g. for environmental studies two processes are important.

One is the tidal mixing which is associated with instantaneous velocities, and the other is the long term convection due to residual and wind driven currents.

To quantify the effect of bottom friction on computed velocities we follow the technique described by Leendetse and Liu (1974) which makes use of spectral analysis. Figs. 6 and 7

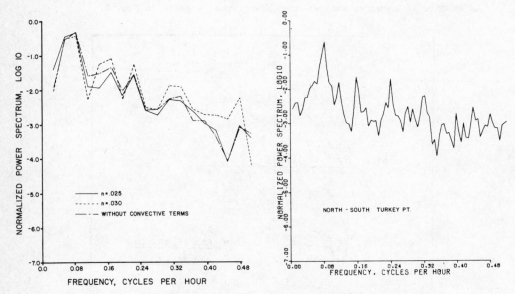

Fig. 6 - Normalised Power Spectra of Computed and Measured
Velocity Components.  Station A-2, Turkey Point

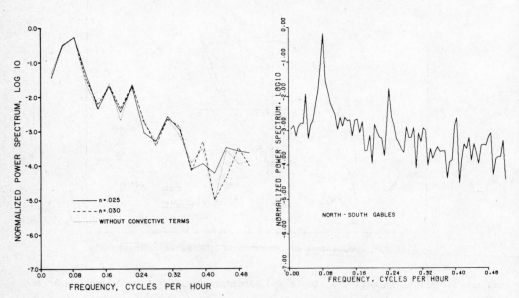

Fig. 7 - Normalised Power Spectra of Computed and Measured
Velocity Components.  Station A-1, Gables

show comparisons between normalised power spectra of the north-south component (east-west components are quite similar) of the velocity at two locations in Biscayne Bay. A total of 256 data points spaced at 500 sec were used to compute the model spectrum, while 2048 points at 600 sec intervals were used for the measured data. This explains the difference in resolution. Both spectra were appropriately smoothed.

For both locations the model spectrum was computed for two values of Mannings n = .025 and .030. In addition a run was made with n = .025 and without the convective acceleration terms in the equations. In the real world the bottom friction, in simplistic terms, causes energy at a basis frequency to disperse into higher frequencies until viscous dissipation eventually transforms the mechanical energy into heat. In a model this process is approximated by a non-linear, usually quadratic friction-law, which tends to emphasize the generation of odd (especially 3rd) harmonics. The convective accelerations are non-linear and thus also contribute to dispersion of energy although favoring even (2nd) harmonics. Fig. 6 shows that an n value of .025 produces excellent agreement between model and data for all significant frequencies. On the other hand, Fig. 7 shows that the model at that location is insensitive to bottom friction and the discrepancy between measured and computed velocities must be explained by the imposed boundary conditions. The different behavior at the two locations can perhaps be explained by the small water depth at Turkey Point and large depth at Gables.

The use of spectral analysis allows better calibration of the friction coefficient and can also be used to verify the validity of a given friction law. With available fast Fourier transform methods this analysis can be made very efficiently.

## Wind Stress Parameterization

Theoretically it should be possible to verify the treatment of an applied wind stress in the model by performing cross-spectral analysis between wind and water velocity. For the present case such attempts were however unsuccessful. A number of possible explanations can be given:

1. The wind induced velocity at a location is not only dependent on the local wind stress but results from the response of the entire water body. This is pursued further in the following section on boundary condition.

2. The wind induced velocities are often very small, at least when vertical uniformity is achieved, and may be obscured by "noise" in the record from large scale turbulence and short period surface waves.

3. The wind field needs to be fairly accurately determined, and it is possible that

> spatial and temporal variations, gusts,
> may be important.

With present technology it has just become possible to investigate the validity of wind stress parameterizations in models, however a carefully planned field monitoring program is required.

For vertically integrated models it is particularly important to evaluate the model response to wind because of the rather restrictive assumption of vertical uniformity and because wind driven currents play a major role in long term convection.

## Boundary Conditions

The precise specification of boundary conditions is of crucial importance in predictive modeling.

For vertically integrated models the proper boundary conditions, when neglecting internal stresses, consist of specifying velocities normal to a boundary, or of specifying the pressure at a boundary. The two conditions are usually prescribed along shorelines and open boundaries respectively.

The pressure can have barotropic, geostrophic and baroclinic components. The barotropic pressure is composed of contributions from long waves, tidal waves, seiches, and wind surges. The sum of barotropic and geostrophic components is directly related to surface elevation while the baroclinic forcing is determined by the density field. At open boundaries the surface elevation and the density should therefore be prescribed.

The treatment of baroclinic boundary conditions depends on the origin of the density variations, however this subject is very complicated and will not be discussed here.

The surface elevation oscillations due to tides can be determined with well-known harmonic analysis techniques, see e.g. Dronkers (1964). The situation in Biscayne Bay is somewhat special because of the multiple (six) openings to adjacent water bodies. Tide gages were installed at all openings and the records analyzed to determine the harmonic constituents. For record lengths of 14 days the variance explained by astronomical tides was consistently between 93-97% of total variance. When using the derived harmonic constituents to forecast surface elevations the explained variance was reduced to approximately 85%. This may be explained by perturbations in the records due to surges on the continental shelf and wind set-up within the bay itself. The maximum difference in predicted and observed tide range was approximately .3 ft. on a 2.0 ft. range or an error of 15%, not uncommon for shallow coastal waters.

For comparison it was found that spectral analysis of the velocity recordings showed that astronomical tide components only explained between 50-75% of the variance depending on location in the bay. The dispersion of energy due to non-linear effects mentioned earlier gives a partial explanation, however in general it is found that non-tidal components make up a significant part of the velocity field.

These non-tidal components are due to wind driving, short surface waves, large scale turbulence, seiches (resonance within a bay), and non-tidal forcing through open boundaries.

Present models cannot handle short surface waves. The occurrence of large scale eddies requires modeling of turbulence and internal stresses and probably should be treated three-dimensionally, see e.g. Leendertse and Liu (1978). Seiches can be treated if a suitable radiation condition can be applied, see Abbott et.al. (1978).

In the following a method is suggested for including the wind driving in the boundary conditions. This takes the form of a set-up (set-down) and a tilt of the water surface. It should be pointed out that it is necessary to include these effects on boundary conditions in addition to applied surface stresses in order to model wind driven flow realistically.

In the present case we have concentrated on determining the slope or relative surface level difference between the open boundaries of Biscayne Bay, since this para-meter is particularly important for the resulting circulation.

The astronomical tide components were subtracted from the records of tide gages C and F, Fig. 1, over the same 2 month period and a new time series was formed by sub-tracting the two residual series. This new time series represent wind induced surface slopes and other non-tidal effects. A cross-correlation was computed for this new series and the north-south or east-west components of wind velocity recorded at Miami International airport. These cross-correlation are shown in Fig. 8. The figure shows that there is no significant correlation between surface slope and east-west component of wind speed while some correlation is found with north-south wind component. The preference towards north-south component of wind is not surprising considering the orientation of the Bay and tide gages used for the analysis. The figure also shows that the surface slope response lags the north-south component of wind speed by 3 hours. At lags of +23 hours and -34 hours the cross-correlation has dropped to a value of $\frac{1}{e}$ of the maximum value and this lag interval is taken as the period over which the surface slope is influenced by the wind speed at zero lag. This information is then used to compute a weighted average slope for each wind recording in the data by averaging the surface slope recordings over the corresponding lag interval. For simplicity a uniform weight was used. The averaged surface slopes for each recorded north-south component of wind speed are then plotted in Fig. 9. Although there is significant scatter in the data points a trend is obviously present. A least squares regression line was fitted to the data and is also shown in Fig. 9. The equation for this line is

$$\Delta y = -.00451 \ x \ -.00047$$

$\Delta y$ = wind induced slope between tide stations C and F

x = wind speed component towards south.

and can be used to include wind induced surface slopes into the prescribed boundary conditions when attempting to model wind driven flow.

Fig. 10 shows predicted lagrangian particle paths for approximately two tidal

CROSS CORRELATION OF C-F AND WIND TOWARDS EAST

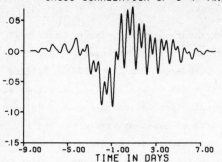

CROSS CORRELATION OF C-F AND WIND TOWARDS SOUTH

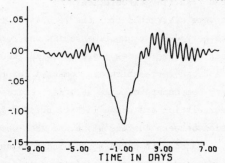

Fig. 8 – Cross Correlation of Surface Level Difference and Wind Components

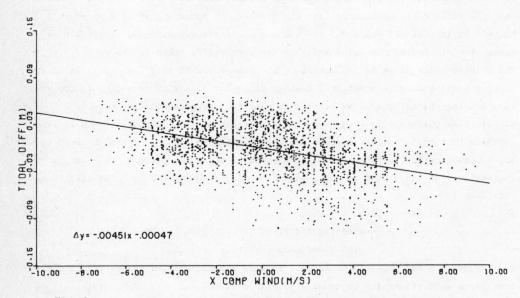

$\Delta y = -.0045|x - .00047$

Fig. 9. – Regression of Wind Speed Component Towards South, x, and Surface
Elevation Difference Between Stations C and F, $\Delta y$

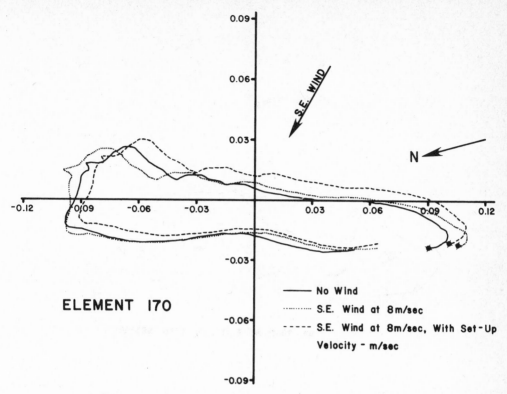

Fig. 11 – Computed Hodographs at Fixed Location in
Biscayne for Tidal Flow with and Without Wind

Summary and Conclusions

Verification of mathematical models is needed to test the validity of underlying assumptions, to determine sensitivity of computed results, and to establish the accuracy with which predictions can be made.

For these purposes it is necessary to be able to quantify differences between computed results and field measurements. The use of conventional spectral analysis techniques seems to be a useful alternative to the usual comparison in the time domain.

In verifying a vertically integrated hydrodynamic model for Biscayne Bay the spectral analysis technique provided a simple way of evaluating bottom friction laws and friction coefficients. From comparison with one current meter record it appears that the quadratic bottom friction law accurately describes the dispersion of energy from a basis frequency to its higher harmonics.

Analysis of surface elevation records showed strong dependence on astronomical tides with typically 85% or better of total variance explained by tidal components.

Similar analysis showed that only 50-75% of the variance in velocity recordings could be directly explained by tides. This higher "noise" level appears to be typical

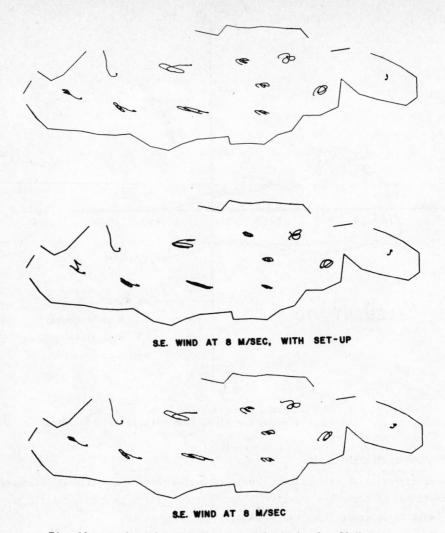

**S.E. WIND AT 8 M/SEC, WITH SET-UP**

**S.E. WIND AT 8 M/SEC**

Fig. 10 - Predicted Lagrangian Particle Paths for 21 Hours

cycles and demonstrates the effect of including the wind induced surface slopes into the boundary condition for the Biscayne Bay model. In Fig. 11 these effects are demonstrated on computed hodographs at a fixed location in the bay. Although it appears that the difference in instantaneous velocities is small it is obvious that the long term average or residual currents can be drastically changed.

for current velocities and sets a limit for the accuracy achievable by deterministic models.

A satisfactory method for verifying wind response predicted by models has not been found. This is due to the difficulty in measuring and identifying the wind induced current, and because of the indirect response resulting from pressure gradients caused by wind set-up of water levels against the shoreline.

Most numerical models are formulated as boundary value problems and to a large extent the boundary conditions naturally will determine the accuracy of computed results. For Biscayne Bay the tidal elevations at open boundaries account for 80-90% of the total variance. An approach is suggested to further reduce the remaining variance and to improve the prediction of wind driven currents by taking wind set-up at open boundaries into account. The resulting corrections to the computed flow field are insignificant when considering instantaneous velocities, but may be important for long term average convection.

### Acknowledgement

This research was supported by NOAA, Office of Sea Grant, Department of Commerce, under Grant No. 04-7-158-44115.

### References

Abbott, M. B., et.al. "On the modeling of short waves in shallow water". J. Hydraulic Research 16, No. 3, 1978.
Dronkers, J. J. "Tidal computations in rivers and coastal waters". North-Holland Publishing, 1964.
Leendertse, J. J., Liu, S. K. "A water-quality simulation model for well-mixed Estuaries and coastal seas : Vol.VI". R-1586-NYC, The New York City Rand Institute, September 1974.
Liu, S. K., Leendertse, J. J. "Multidimensional numerical modeling of estuaries and coastal seas". Advances in Hydroscience, Vol. 11, Academic Press, 1978.
Wang, J. D. "Real-time flow in unstratified shallow water". J. Waterway, Port, Coastal and Ocean Division, ASCE, WW1, February 1978.

Mathematical Modelling of tidal flats : a few remarks

H. Holsters

Abstract

Comments are presented on a few computational aspects of mathematical-hydrodynamical
modelling of the influence of tidal flats on the regime of an estuary.
The numerical integration of the equations for quasi horizontal two-dimensional flow
is effectuated on a square computational grid using an explicit time-space staggered
finite difference method. Two ways of space-staggering of the velocity points versus
the elevation points are considered.

## 1. Introduction

The Scheldt estuary is part of the Scheldt-Rhine-Maas Delta situated in the south-
west of the Netherlands. About one third of the total area of this estuary is situated between the high- and low-waterlines and consists of sandbars and salt marshes.
The sandbars are located in between the channels and, owing to their convex and generally elongated shape, are readily accessible to the rising tide and easily drai-
ned during ebb. On the salt marshes the situation is much more complex. Here the flood
has to gain access through a very complicated system of principal and secundary creeks,
the smallest of which sometimes end in low and wide depressions. At strong spring-
tides and during storm surges the area is completely covered at high water and the
motion is two-dimensional on the whole area. During draw-down the flow pattern again
degenerates into a more or less coherent system of one-dimensional channels until
the water either recedes completely or is left behind in some of the depressions.
In the context of a general study of this estuary and the adjacent area of the North
Sea a series of two-dimensional hydro-dynamical test computations is currently being
carried out using two different versions (to be referred to as the "lateral" and the
"diagonal" methods) of an explicit staggered finite difference technique similar to
the well-known HN-method [3] , [4] . The aim of the present paper is to comment on some
computational aspects connected with the modelling of tidal flats.

## 2. Lateral Method

The first version is based on the traditional form [4] of the HN-method. The elevation h is calculated (or prescribed on the open boundary) at the nodes of a square
grid with cartesian coordinates $x = i\Delta x$, $y = j\Delta y$, and $\Delta x = \Delta y = \Delta s$.
The depth-mean x and y velocity components, u and v, are located on the sides (i.e.
laterally) of the grid squares, with grid-coordinates (i+1/2, j) and (i, j+1/2), res-

pectively. By virtue of the zero normal flow condition either u or v vanish at the closed boundary, which is represented by a polygonal contour piecemeal parallel to the grid directions and passing through (or changing direction in) the centre of the grid squares.

At the elevation points the bottom level $z_b$ is defined by means of the actual depth below datum. At the velocity points the bottom level is obtained by linear interpolation between the bottom levels of the adjacent elevation points.

With respect to the small depths obtaining on the flats, and in order to avoid the numerical instability which might occur if the frictional resistance is simulated in a purely explicit way (cf [5], [7]), a straightforward yet semi-implicit formulation (cf [1],[5]) is chosen as indicated in Appendix I.

Another difficulty, connected with the schematization of variable bottom shape, is encountered, especially on the salt marshes, because the sub-gridsize details of the complex bottom morphology (e.g. the narrow creeks which take care of a non-negligible fraction of the outflow during draw-down) are smoothed out through depth averaging. The schematization can be improved by introducing a modified Chezy coefficient [2] or a modified friction factor [6]. In the testcomputations a similar improvement is obtained by replacing the flow depth D in the denominator of the friction term with D + D', where D' is an empirical parameter and in fact corresponds to a particular case of the ad hoc formula presented in [6].

During ebb, and by virtue of the mass conservation principle, the total volume flowing during a full time step from a grid block surrounding the elevation point under consideration normally equals the product of the block area and the change of water level. "Falling dry" occurs when not enough water is available to balance the numerical net outflow, and a negative depth will be found. As negative depth is not allowed the water level has to be taken equal to the bottom level $z_b$, but this does not alter the fact that more water is numerically being restituted to the main flow than was stored during flood at the elevation point under consideration.

In order to avoid such errors a marginal thickness d is introduced equivalent to a boundary layer of stagnant water, and the elevation point is "inactivated" (i.e. taken out of the computation) as soon as the local water level is lower than the prescribed minimum level $h_{min} = d + z_b$. The elevation point under consideration will be re-activated when, at run-up, the water level in its vicinity exceeds the minimum level $h_{min}$.

Normally a newly inactivated point will remain inactive after drawn-down until the next flood period, and a newly activated point will remain active after run-up until the next ebb period. However, if the marginal thickness d is chosen too small, at certain points situated near the steeper edges of the tidal flats, a certain number of successive draw-down/run-up cycles may occur, leading sometimes to non-negligible parasitic outflow. These numerical perturbations can be suppressed either by increasing the marginal thickness d (assuming this to be possible without significant reduction of the storage capacity of the tidal flats), or by refining the grid.

## 3. Diagonal method

In the second version of the HN-method the same geometrical configuration of elevation points open boundaries and closed boundaries is used as in the lateral method. However the coordinate system is rotated through a 45 degree angle relative to the grid lines. Consequently the components u and v now are parallel to the <u>diagonals</u> of the grid and are both calculated at the centres of the grid squares formed by the elevation points (see also figure 1, Appendix II ).

The interval $\Delta s = \Delta x = \Delta y$ now being $\sqrt{2}$ times larger than in the lateral scheme, the time step $\Delta t$ can be increased in the same proportion by virtue of the C.F.L. stability criterium.

With respect to the zero normal flow requirement usually the no-slip condition u = v = 0 is adopted along the closed boundaries, or the same effect is obtained by assuming zero depth. However, this results in the introduction, along the coastline, of a boundary zone of the order of half the mesh width, and this leads to a prohibitive reduction of the available cross-sectional areas in an estuary with tidal marshes and relatively narrow low-water channels and by-channels.

On the other hand, satisfactory results may be obtained in this case when using the set of boundary conditions indicated in appendix II : u=0 or v=0 at "re-entrant" angles (90°); u = v = 0 at "outside" angles (270°); u + v = 0 or u - v = 0 along the straight stretches.

With respect to the bathymetric schematization the actual bottom level is used at the velocity points. At the elevation points the bottom level is needed only on the tidal flats, and the actual bottom level is used, or (in the case of very irregular bottom shape) some weighted average of the actual level and the mean level of the adjacent velocity points.

The features connected with the existence of tidal flats are basically the same as in the lateral method. However, their implementation, as far as the draw-down/run-up conditions are concerned, is less simple (as compared with the lateral method) because the schematization involves more complex geometrical and topographical relations between the elevation and the velocity points.

For instance, the bottom level at a velocity point may be higher than the bottom levels of two or more of the four adjacent elevation points, whereas this is impossible in the lateral scheme where, as indicated in section 2, the bottom levels of each velocity point and its two neighbouring elevation points are collinear. An obvious consequence of this more complex situation is that the existence of a spatial elevation gradient does not any longer automatically imply the existence of a positive flow depth.

Another, less obvious, consequence is that a distinction should be made between the "wet" elevation points situated inside an impounded area and those situated in the main flow. Only elevation points of the latter category may be used for the implementation of the run-up condition.

## 4. Conclusion

The modelling of tidal flats, based on the traditional (or "lateral") form of the HN-method, as presented in section 2 differs only on minor points from existing explicit methods, for instance [1], [6].
The less frequently used "diagonal" form of the HN-method can also be adapted for the modelling of tidal flats, as indicated in section 3.
Computations on a relatively coarse schematization (mesh width 1000 meter)of the lower reaches of the Scheldt estuary having resulted for both methods in a satisfactory agreement with the available observational information, the comparative computations will be extended, with a finer grid, to the whole estuary.

## Appendix I : Frictional resistance in shallow water

A purely explicit formulation of the frictional resistance in a shallow tidal river may cause computational instability, and semi-implicit[5] or implicit [7] expressions are in order. A straightforward yet semi-implicit formulation is obtained in the following way.
Consider the simplified dynamic equation for one-dimensional flow :

$$u_t \ + \ gh_x \ + \ ru \ |u| \ / \ D = \ 0 \qquad\qquad (1)$$

where : u = cross-sectional mean velocity; h = water level; D = mean water depth = cross-sectional area / surface width ; r = friction factor; g = acceleration of gravity.
By using finite differences on a staggered space-time grid, equation (1) can be transformed into :

$$U' \ = \ U \ - \ 2r\Delta t U' \ |U| \ /D \ + \ (h_w - h_e) \ g\Delta t \ /\Delta x \qquad (2)$$

where : $\Delta t$ = time step; $\Delta x$ = space interval; U' = u (t+$\Delta$t,x); U' = u (t-$\Delta$t,x); D = D(t,x); $h_w$ = h (t, x-$\Delta$x); $h_e$ = h(t,x+$\Delta$x).

Formally this is an implicit expression with respect to the velocity, but it can easily be rewritten as a straightforward expression:

$$U' \ = \ ( \ U + \ (h_w - h_e) \ g\Delta t \ /\Delta x \ ) \ R \qquad (3)$$

with $\qquad R = (1 + 2r\Delta t \ |U| \ /D)^{-1} \qquad\qquad (4)$

In the two-dimensional case a similar resistance coefficient can be used to calculate the velocity components u and v by replacing U in formula (4) with $(u^2 + v^2)^{1/2}$.

Appendix II : Closed-boundary conditions of the "diagonal method"

In order to show how the condition of zero normal flow along a closed boundary may
be implemented in a pragmatic way when using the diagonal method, let us consider
figure 1 where a shoreline is represented consisting of two straight stretches PQ
and QR.

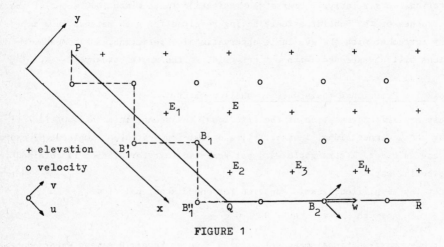

**FIGURE 1**

The u- and v- directions forming a 45° angle with the grid lines the zero normal
flow condition along the horizontal boundary QR requires;

$$u\cos 45° - v\sin 45° = (u-v)\cos45° = 0.$$

Furthermore a longitudinal velocity equal to $(u+v)\cos45°$ can be calculated in a
point such as $B_2$ by means of the longitudinal gradient generated by the water levels
of the elevation points $E_3$ and $E_4$. Likewise, it can be shown that the zero normal
flow condition along a vertical boundary is $u + v = 0$, and that both components
of a longitudinal velocity can be calculated if two suitable located elevation
points are available.

Near the diagonal boundary PQ a longitudinal component u can be computed directly
in a point such as $B_1$ using the gradient generated by the elevation points $E_1$ and
$E_2$. The normal component v cannot be calculated because only one elevation point
(in casu E) is available in the y-direction and thus the most likely value v=0 is
assumed. Likewise, in the case of a diagonal boundary parallel to the y-direction
the zero normal flow condition corresponds to u=0, and a longitudinal component v
can be computed if two suitably located elevation points are available.

As shown on figure 1 the schematized boundary representing the shoreline is a step-

line. In outside points such as $B_1'$ the velocity components cannot be calculated
(only one elevation point, in casu $E_1$, is available) and it is assumed that u and v
may be taken equal to zero.

In the general case of a closed boundary schematized by an irregular succession of
step-lines and straight lines the boundary conditions can be summarized as follows :

a) at a re-entrant point (90° top-angle, measured overland); u=0 or v=0.

b) at an outside point (270° top angle, measured overland); u=v=0.

c) at any other boundary point; u+v = 0 or u-v = 0.

With respect to a), and referring to figure 1, it should be noted that at the re-
entrant angle $B_1' - B_1 - B_1''$ the boundary condition is not rigourously consistant with
zero normal flow. In fact, across the segment $B_1'B_1$ a non-zero outflow equal to
$(uMD_1 \cos 45°)/2$ (with M=mesh width and $D_1$ = depth in $B_1$) exists. But as this outflow
is exactly compensated by an inflow of the same magnitude across the segment $B_1 B_1''$,
the total net flow across the contour $B_1'B_1B_1''$ is zero.

## References

1. Backhaus J., "Zur Hydro-dynamik im Flachwassergebiet. Ein numerisches Modell",
   Deutsches Hydrographisches Zeitschrift, 1976.

2. Dronkers J.J., "The schematization for tidal computations in case of variable
   bottom shape", 13th Coastal Engineering Conference, Vancouver, 1972.

3. Hansen W., "Theorie zur Errechnung der Wasserstanden und der Strömungen in Rand-
   meeren nebst Anwendungen", Tellus 1956.

4. Hansen W., "Hydro-dynamical methodes applied to oceanographic problems", Mit-
   teilungen des Institutes für Meereskunde der Universität Hamburg, 1972.

5. Holsters H., "Remarques sur la stabilité dans les calculs de marée", Proceedings
   of the Symposium on Mathematical-Hydrodynamical Methods of Physical Oceanography,
   Hamburg, 1961.

6. Ramming H.G., "Numerical investigations of the influence of Coastal structures
   upon the dynamic off-shore process by application of a nested tidal model", Elsevier Oceanography Series, No 23, 1978.

7. Vreugdenhil C.B., "De invloed van de wrijvingsterm op de stabiliteit van differen-
   tiemethoden voor hydraulische problemen", De ingenieur, 1966.

# ON STORM SURGE PHENOMENA

M.Laska

Institute of Meteorology and Water Management
Gdynia, ul.Waszyngtona 42, Poland

## Introduction

The reader is aware of the human and material losses due to storm surges.
Therefore, there is no need to recall in details the tragedy of Holland
in 1953, Japan in 1959, Germany in 1962, or all the other floodings
which took place in various countries all over the world in the past or
most recent years.

There are many excelent reviews and papers published on storm surges.
A sample of them is given in the references, and worth to say, one of
the most comprehensive list of references on this problem was published
already in 1954 by, then called, International Association of Physical
Oceanography.

A lot of discussion is going on, and Conferences or Symposia were and
are being held on this topic in order to exchange experiences and views
on the storm surge problem, to give ways how to achieve the common goal
i.e. how and by what means to cope with the storm surge phenomena.

The paper presented is not ment, neither to give the right remedy for
the storm surge problem, nor to present the most sophisticated and reliable predictive model for this purpose. It is ment only to describe,
in general, the problem in concern and to repeat once more and again
the urgent need for cooperative studies and efforts in the field of
storm surge phenomena in order to avoid in the future any human and
material losses if possible.

The simple, general definition of a storm surge reads: "storm surge is
an abnormal rise or fall of water surface above or below its mean level
induced by meteorological driving forces". Of course this definition is
rather coarse and does not describe the phenomenon in more detail.
A better description is always available if one takes into consideration
all the physical components involved in a generation of a surge, i.e.
the driving forces, the physical characteristics of the sea and the sea
area where the surge takes place, the physical characteristics of the

surge itself and all the additional facts interacting with the phenomena in concern.

The dangerousity of a surge rises always in respect to the phasing of
all the components taking part in the generation of a surge. Although
there are many, sometimes even difficult to describe, physical components involved in the "effectivenes" of a surge, but one thing remains
obvious that in general it is not the surge itself that causes losses
but the long lasting dynamical activities of waves riding atop the ele-
vated sea surface, creates the inundation potential.

Many coastal areas are particularly vulnerable to storm surges and, therefore, there is a great need and a good deal of attention has to be
attracted as well as much effort has to be spend in developing models
to cope with the prediction of the dangerous phenomena.

There is a wide range of models for this purpose but, true to say, none
of them is neither fully acceptable from the engineering point of view
nor totally reliable to deal practically with the problem. Of course
some of the models developed are succesfully used for general and partial prediction but sofar there is no such a model which can be practically and moreover universally used for prediction in any sea area vulnerable to storm surges.

Storm surge modelling

The up to date existing models for water circulation in general, and
storm surge phenomena in particular, are classified into various groups.
For example (5) gives a classification in respect to the modelling approach; having an empirical approach, semi-empirical semi-theoretical
approach and theoretical approach. Of course all classifications, what
ever they are, are always subjective and left open for amendments, depending on the technique used and the point of view the modeler is going to take giving his own classification.
The known modelling technique of storm surges can be classified into
three groups. Let us put in the first group the empirical and hydraulic
models, in the second the analitical and statistical models, and in the
last the numerical and analoque models. The grouping can be set of course quite differently. For instance statistical and numerical models
can be tied together by power spectrum analysis, or to the whole classification graphical methods, say, method of characteristics can be
supplemented.

As seen the grouping given above is not a final one, it is subjective

as well. It is given only for the sake of an insight into the recent
modelling technique, and in order to characterize the pros and contras
of each method.

The oldest and most commonly used are the empirical and hydraulic models. Models of this group give satisfactory results as far as the simple prognostic relations (empirical models) and hydrographic analysis
(hydraulic models) are concerned. Generally an empirical model is a set
of regretion equations, relating surge to pressure gradients, the latter evaluated statistically from tide gauges and weather charts in
storm conditions. The results obtained describe the average characteristics of the surge, and as the equations are applicable to particular
tidal station or sea area only, the applicability of this method is
rather restricted. Nevertheless, it is sofar the only method in operational use. Worth to mention that no type of mathematical model for water circulation in coastal areas can do without empirical information
about the exchange of momentum and mass.

Good practical information can give the hydraulic modelling technique,
especially if connected, as recently is widely observed, with a mathematical model, giving thus another grouping namely the hybrid models.

A superior role, as far as the basic research is concerned, play in the
second group the analytical modelling technique. But as their mathematical solution is based on many assumptions and simplifications and as
they are restricted, in general, to ideal conditions and imaginary reservoirs, more practical attention is given to statistical models.
The latter suffers, of course, from the pausity of hydro-meteorological
data, giving in addition only a statistical quantitative characteristics of storm surges without any chances to describe the physics of the
phenomena in concern. Most of the restrictions mentionned can be avoided when using the third group of models, especially the numerical models.

The numerical modelling technique, before it was taken into the field
of physical oceanography (6) was already applied for meteorological
progostic purposes.

The numerical technique in essence employs the finite difference method to solve the governing hydrodynamic-differential equations, replacing the derivatives by difference approximations. Thanks to the vast
availability of fast computers this technique is now widely used in the
field of dynamical oceanography.
For storm surges this technique in one, two and recently in three di-

mensional models, was applied by many researchers showing its great validity for prediction purposes, but true to say, sofar it is not in full operational use. One of the main obstacles here are not the tidious and time consuming numerical calculations for this is computerized, but the lack of sufficient, fully reliable and in time advanced input data.

The accuracy of the results of numerical modelling depends of course, upon the numerical method used, whether it will be explicit, implicit, semi-explicit, semi-implicit or a combined scheme, and on the boundary conditions as well as on the size of the numerical grid. The size of the grid and time step must be chosen carefully, and great care should be taken on numerical viscosity and/or diffusivity which can disturb the accuracy of the method used.

The coarseness of the model grids and the use of nested grids, especially in three dimensional models, which can take into account the rapidly developing knowledge of turbulence, has attracted a good deal of attention and it is hoped that this will greatly improve the storm surge forecasting technique. Great care as well has to be taken to the verification of the model deviced. This assures that the solution procedure is correct, the formulation is cosistent and all the assumptions made are reasonable.

Any model results must be compared with actual field data, but the existing field monitoring program do not always satisfy these minimal requirements, partly due to inadequate technology.

In practice there are two types of numerical models, namely continental shelf models,e.g.(4,7,9,13) and bay-estuary or basin models,e.g.(1,2,6, 10,12,16,22,24), fast all focused on hindcast phenomena.
Shelf models compute surges in space and time in the open sea and open coast, whereas bay and basin models are restricted to smaller areas and are influenced by small-scale features of the coast -here additional physics has to be taken into account then in shelf models.

There is no need to emphasize that the state of art regarding the methodology of theoretical prediction of storm surges is very advanced, but as said before, no uniform theory exists sofar which could be "sine qua non" applied for practical forecast of these phenomena. One can hope that the newly, in dynamical oceanography emerging method, namely the finite element technique,(characterized in one of the papers here) will fill one of the gaps in storm surge forecast. Although the hindcasting is very promising, still a lot has to be done untill a fully acceptable model will be in hand for the "ad hoc" storm surge forecast purposes.

## Generation of storm surge phenomena

It is obious that hydrodynamical processes of storm surge development
can be clearly described if the forces which generate surges and the
associated motions of the sea will be adequately evaluated.

In general, the principal factors and causes involved in the generation
of storm surge phenomena can be described as given hereafter.

a. The effect of wind stress on the sea surface.

The wind acting over the sea surface piles up the water. For this phenomena the name of wind set-up is often used. The set-up of the water
caused by wind stress is generally thought to consist of two components.
The first component is the set-up of water due to the onshore wind.
Here the generated slope of the sea water is directly proportional to
the wind stress and inversely proportional to the depth of the water.
The effect of the wind blowing parallel to the shore, and generating
to the latter parallel current, is the second component.
The wind stress on the surface, or better the tangential stress at the
sea surface due to the wind action is generally assumed to be proportional to the square of the wind speed. The product of the atmospheric
density and a parameter called the drag coefficient expresses the characterized relationship.
The oceanographic literature describes various methods of measuring the
drag coefficient. It can be done by direct measurments from the wind
profile existing above the sea surface or indirectly, although less reliable, from the slope of the sea surface caused by steady wind conditions,(22 ). It is already customary that for wind speeds smaller than
15m/sec the drag coefficient is calculated from the wind profile, whereas for winds over 15m/sec up to 3om/sec the measurments from the water slope are considered.

b. The effect of atmospheric pressure acting on the sea surface.

A depression moving into the sea area with its atmospheric pressure
acting normal to the sea surface brings a fall in the pressure causing
in consequence a rise of water level. Leaving the area the pressure
rises and the water level falls. Therefore, a rise followed by a drop
of water level occurs during a passage of a depression. A reduction of
atmospheric pressure in storm areas causes mostly an increase of sea
level; as a rule in accordance with the assumption of the statical law.
This effect is generally called the inverted barometer effect, causing,
in hydrostatic considerations, a rise of sea level of approximately

1 cm by a drop in atmospheric pressure of 1 mb. This provides an approximate value of a surge due to a pressure drop in a cyclone.
In general the atmospheric pressure has a smaller effect on the rise of sea level than the wind stress effect. This is of course true for shallow water areas where surges are caused mainly by wind, whereas in deep water (where surges occur as well, but are not evaluated due to their small practical importance) the atmospheric pressure effect predominates.

c. The effect of water transport due to waves and swells especially in shallow water areas.

The tractive force on the sea surface exerted by the wind causes a motion of the water masses. As a result the water is transported along the direction of the wind as long as the external generating source and unless the water flow is damped out by other forces, e,g. friction.
When the net transport of the water reaches the coast the water level rises. Various investigators agree (19 ) that the slope of water surface caused by waves (in our case long waves) is directly proportional to the gradient of variance in water level due to waves (proportional to wave height squared), and inversely proportional to the depth.
Of course the effect of Earth's rotation in both cases plays an important role. It will act in changing the direction of the flow of the current and the resulting acceleration will greatly affect, depending on the dimensions of the reservior in concern, the storm surge phenomena. The acceleration, due to Coriolis force, as known, is to the right in the northern hemisphere and to the left in the southern hemisphere.

d. The modifying effects of storm surges.

On the final results of storm surge phenomena, besides the external forces, especially wind and pressure effects, and the modifying effect of the Coriolis force, an important role play in this case the existing physical characteristics of the basin in concern.
The geometry of the basin can cause divergence or convergence of the surge. This depends on the shoreline orientation, depth distribution and existing hydraulic structures. The surge at any location can be modified due to local conditions. These are due to change in some way, especially as far as depth distribution is concerned according to the existance of astronomical tides. In tidal areas the surges interact nonlineary with astronomical tide especially when tidal ranges are significantly large compared to the effective mean water depth.

The tide can amplify the water set-up due to the surge to a great extent. This value will be smaller by low water periods but greater at high water periods. The interaction of the surge with tidal currents, especially in basins with high values of the latter has also to be taken into account . In tidal estuaries besides the existing tide, the river discharges can reinforce and augment the built up of the surge. The most critical situation occurs of course when the peaks of the surge, tide and river discharge are in phase.

Fig.1 describes schematically the causes of a storm surge generation and the way how to evaluate the phenomena in concern.

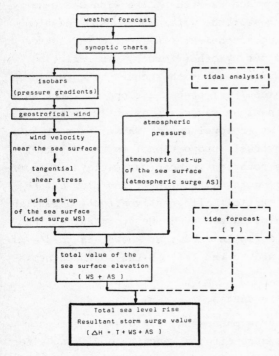

As it is seen from the flow-diagram given the storm surge prediction problem includes two separate disciplines; meteorology and oceanography. The two components are mutually dependant and exist in a well developed interaction. It is obvious that the outcome of the oceanographic component depends on the meteorological part; i.e. the space-time field of the driving forces on the sea surface has to be correctly evaluated. Needless to say that a good designed oceanographic model, even with the use of highly sophisticated computer, will be useless if the meteorological model, being the feeding part of the whole design, is poorly described. True to say here -vice versa as well.

Fig. 1 Flow diagram of the storm surge causes and the way of its calculation

The tide, in basins of astronomical tides, adds complication to the surge prediction, for there exists a strong nonlinear interaction between these two phenomena, especially in shallow water areas, (1). The phasing problem of various conditions is one of the main factor in storm surge modelling. It is clear that the resultant effect of the storm surge depends totally on; the driving forces, the existing water level (whether high or low water) the tidal wave and the geometry of the basin in concern. If there is the phase agreement between high water level, tidal and surge wave, long duration of the driving forces and moreover the area is vulnerable and susceptible to surges, as was the fact in Holland in 1953, the outcome leads to most dangerous effects.

An example of numerical calculation of water circulation in the Baltic

A hydrodynamical-numerical (H-N) model of wind-driven and river inflow steady circulation in the Baltic Sea was lately deviced ( 12 ). This model is applied here for a general simulation of a storm surge phenomenon.

The basic set of equation is given in the form:

$$\frac{\partial U}{\partial t} - \Omega V - A \frac{\partial^2 U}{\partial z^2} + g \frac{\partial \zeta}{\partial x} + \frac{1}{\rho_0} \frac{\partial p_a}{\partial x} = 0$$

$$\frac{\partial V}{\partial t} + \Omega U - A \frac{\partial^2 V}{\partial z^2} + g \frac{\partial \zeta}{\partial y} + \frac{1}{\rho_0} \frac{\partial p_a}{\partial y} = 0$$

$$\frac{\partial U}{\partial x} + \frac{\partial V}{\partial y} + \frac{\partial W}{\partial z} = 0$$

where t = 0 and U = V = W = = 0.

At the free surface (z = $\zeta$ ) the following boundary conditions were taken;

$$\rho_0 A \frac{\partial U}{\partial z} = T_x{}^s; \qquad \rho_0 A \frac{\partial V}{\partial z} = T_y{}^s$$

$$W_\zeta = U_\zeta \frac{\partial \zeta}{\partial x} + V_\zeta \frac{\partial \zeta}{\partial y} + \frac{\partial \zeta}{\partial t}$$

whereas at the bottom Z = - H;

$$\rho_0 A \frac{\partial U}{\partial z} = T_x{}^b = RMx; \qquad \rho_0 A \frac{\partial V}{\partial z} = T_y{}^b = RMy$$

$$W_H = -U_H \frac{\partial H}{\partial x} - V_H \frac{\partial H}{\partial y}$$

Intergrating the equations in vertical from the bottom to the surface we obtain the following set of equations for mass transport and water elevation in the form:

$$\frac{\partial Mx}{\partial t} - \Omega My = Tx^s - RMx - \rho_o g \ (H+\zeta) \ \frac{\partial \zeta}{\partial x} - (H+\zeta) \ \frac{\partial P_a}{\partial x}$$

$$\frac{\partial My}{\partial t} + \Omega Mx = Ty^s - RMy - \rho_o g \ (H+\zeta) \ \frac{\partial \zeta}{\partial y} - (H+\zeta) \ \frac{\partial P_a}{\partial y}$$

$$\frac{\partial \zeta}{\partial t} = - \frac{1}{\rho_o} \left[ \frac{\partial Mx}{\partial x} + \frac{\partial My}{\partial y} \right]$$

Where:

$\quad$ u,v,w $\qquad\qquad$ -components of current velocity along x,y,z axes
$\qquad\qquad\qquad\qquad$ respectively (the origin of the coordinates is
$\qquad\qquad\qquad\qquad$ at the sea surface, x -points to the East,
$\qquad\qquad\qquad\qquad$ y -to the North, and z -vertical upwards)

$\quad$ $\zeta$ $\qquad\qquad\qquad$ -sea level variation from the undisturbed surface

$\quad$ A $\qquad\qquad\qquad$ -eddy viscosity coefficient

$\quad$ $\rho_o$ $\qquad\qquad\qquad$ -density taken as a constant value

$\quad$ $\Omega$ $\qquad\qquad\qquad$ -Coriolis parameter

$\quad$ g $\qquad\qquad\qquad$ -Earth gravity acceleration

$\quad$ $T_x$, $T_y$ $\qquad\qquad$ -components of the wind stress at the sea surface

$\quad$ R $\qquad\qquad\qquad$ -coefficient of the bottom stress

$$M_x = \int_{-H}^{0} u \ dz$$

$\qquad\qquad\qquad\qquad$ -components of mass transport,
$\qquad\qquad\qquad\qquad$ along x and y axes respectively

$$M_y = \int_{-H}^{0} v \ dz$$

From the given set of equations we can go over to the mean velocities
writing the equations as follows:

$$\bar{U} = \frac{1}{H + \zeta} \qquad\qquad \int_{-H}^{\zeta} u \ dz = \frac{M_x}{H + \zeta}$$

$$\bar{V} = \frac{1}{H + \zeta} \qquad\qquad \int_{-H}^{\zeta} v \ dz = \frac{M_y}{H + \zeta}$$

The equations of motion and continuity are written in the form:

$$\frac{\partial \overline{U}}{\partial t} - \Omega \overline{V} = \frac{T_x^s}{H+\zeta} - R\overline{U} - \rho_o g \frac{\partial \zeta}{\partial x} - \frac{\partial P_a}{\partial x}$$

$$\frac{\partial \overline{V}}{\partial t} + \Omega \overline{U} = \frac{T_y^s}{H+\zeta} - R\overline{V} - \rho_o g \frac{\partial \zeta}{\partial y} - \frac{\partial P_a}{\partial y}$$

$$\frac{\partial \zeta}{\partial t} = - \frac{1}{\rho_o} \frac{\partial[(H+\zeta)\overline{U}]}{\partial x} + \frac{\partial[(H+\zeta)\overline{V}]}{\partial y}$$

Initial boundary conditions can be set as; for mass transport $M_x = M_y = 0$, for mean velocities $\overline{U} = \overline{V} = 0$, and for sea level $\zeta = 0$.
The final set of equations, according to the numerical grid of (12 ), is written now in the finite-difference form:

$$M_{x_{m+1,n}}^{t+\tau} = M_{x_{m+1,n}}^{t-\tau} (1-2R\tau)+2\Omega\tau\overline{M}y - \frac{g\tau}{h} H_{m+1,n}(\zeta_{m+2,n}^t - \zeta_{m,n}^t)+$$

$$+ T_{x_{m+1,n}} - \frac{\tau}{n}H_{m+1,n}(P_{a_{m+2,n}} - P_{a_{m,n}})$$

$$M_{y_{m,n+1}}^{t+\tau} = M_{y_{m,n+1}}^{t-\tau} (1-2R\tau) - 2\Omega\tau\overline{M}_x - \frac{g\tau}{h} H_{m,n+1}(\zeta_{m,n+2}^t - \zeta_{m,n}^t) +$$

$$+T_{y_{m,n+1}} - \frac{\tau}{n} H_{m,n+1}(P_{a_{m,n+2}} - P_{a_{m,n}})$$

(note: H = H+$\zeta$)

$$\zeta_{m,n}^{t+2\tau} = \zeta_{m,n}^t - \frac{\tau}{\rho_o h} (M_{x_{m+1,n}}^{t+\tau} - M_{x_{m-1,n}}^{t+\tau} + M_{y_{m,n+1}}^{t+\tau} - M_{y_{m,n-1}}^{t+\tau} )$$

In the equation of continuity we can write the M-values as:

$$M_{x_{m+1,n}}^{t+\tau} = (H_{m+1,n} + \zeta_{m+1,n}^{t+\tau})\overline{U}_{m+1,n}^{t+\tau}\cdots$$

$$M_{y_{m,n+1}}^{t+\tau} = (H_{m,n+1} + \zeta_{m,n+1}^{t+\tau})\overline{V}_{m+1,n}^{t+\tau}\cdots$$

and the means are evaluated as follows:

$$\overline{M}_x = 0.25(M_{x_{m+1,n+2}}^{t-\tau} + M_{x_{m-1,n+2}}^{t-\tau} + M_{x_{m-1,n}}^{t-\tau} + M_{x_{m+1,n}}^{t-\tau} )$$

$$\overline{M}_y = 0.25 (M^{t+\tau}_{Y_{m+2,n+1}} + M^{t-\tau}_{Y_{m+2,n-1}} + M^{t-\tau}_{Y_{m,n+1}} + M^{t-\tau}_{Y_{m,n-1}})$$

$\tau, h$ - time and space steps of numerical grid.

A first order approximation in time and second in space, for the differential operators, is given in the above written scheme. Numerical stability conditions in the grid, for the case of shortest waves, were given ( 12 ) as follows:

$$\tau \leq \frac{1}{R} \; ; \; \tau \leq \frac{1}{2R}$$

$$\tau \leq -\frac{Rh^2}{4gH} + \frac{h}{\sqrt{gH}}$$

The latter expresses, in a frictionless flow $R \approx 0$, the Courant-Fridrichs-Levy criterion.

The results obtained (some are shown in figures 2 and 3) proved the scheme to be stable for all wave lengths, however, convergence to the steady state was very slow in time; therefore, in order to speed it up the authors ( 12 ) used a smoothing operator in the form:

$$B\Delta M_x; \; B\Delta M_y$$

where $\Delta = \frac{\partial^2}{\partial x^2} + \frac{\partial^2}{\partial y^2}$ and

$B = 1o^7$ is an analogue of the lateral eddy viscosity

The calculations were performed with a time step $\tau = 6o$ sec., and the space step of $h = 5$ Nm. Coriolis parameter $\Omega$ was equal to $1.2368 \cdot 1o^{-4}$

The results obtained (fields of mass transport, surface currents, and water level distribution) indicate a strong coupling with the wind field, Earth's rotation, and bottom topography.

As far as the river runoff to the Baltic basin is concerned, the results obtained indicated a slight influence, in coastal waters only, in the wind-driven circulation, proving that in macro-scale models of wind-driven circulation ( 12 ) the river inflow to the Baltic can be neglected.

As the reader is always faced with a sample of storm surge hindcasting diagrams, where computed and observed values are compared, other kind of results showing the apllicability of HN technique is given for a change.

189

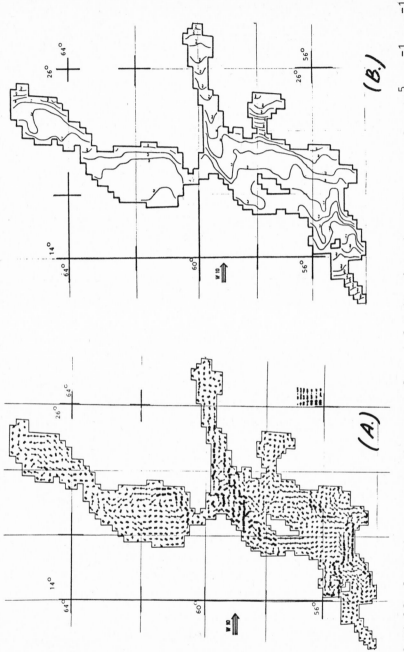

Fig. 3 Field of stationary mass transport for wind W 10 m/sec (mass trans. value $10^5$ gcm$^{-1}$ sec$^{-1}$) — A.
Stationary water level for wind W 10 m/sec (isolines every 2.5 cm) — B. /acc. to (12)/.

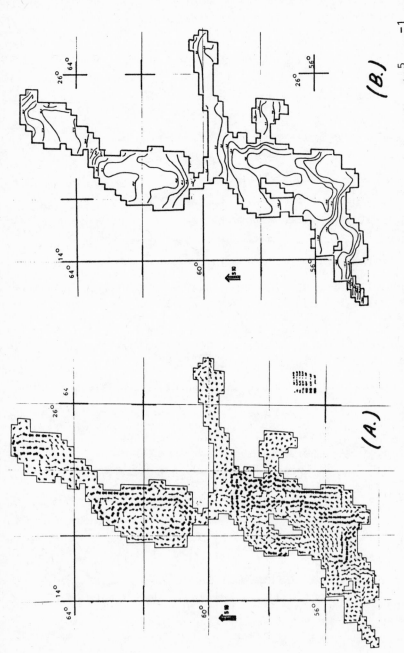

Fig. 2   Field of stationary mass transport for Wind S 10 m/sec (mass trans. value $10^5$ gcm$^{-1}$ sec$^{-1}$) — A.
Stationary water level for wind S 10 m/sec (isolines every 2.5 cm) — B. /acc. to (12)/.

References

(1) Banks,J.E.,1974: A mathematical model of a river-shallow sea sys-
     tem used to investigate tide, surge and their interaction
     in the Thames-Southern North Sea region. Philos.Trans. R.
     Soc. (A),275.

(2) Duun-Christensen,J.T.,1975: The representation of the surface pres-
     sure field in a two-dimensional hydrodynamic numeric model
     for the North Sea, the Skagerrak and the Kattegat.
     Dt. hydrogr. Z.,28.

(3) Fischer,G.,1959: Ein numerisches Verfahren zur Errechnung von
     Windstau und Gezeiten in Randmeeren. Tellus, 11.

(4) Flather,R.A. and Davies,A.M.,1976: Note on a preliminary scheme
     for storm surge prediction using numerical models.
     Quart.J.R.Meteorol.Soc. 102.

(5) Groen,P. and Groves,G.W.,1962: Surges, -in The Sea. Ideas and ob-
     servations on progress in the study of the seas.

(6) Hansen,W.,1966: Die reproduktion der Bewegung vorgänge im Meer mit
     Hilfe hydrodynamisch-numerischer Verfarhen. Mitt. Inst.
     Meereskunde der Uniwersität Hamburg.

(7) Harris,D.L.,1963: Characteristics of the hurricane storm surge.
     U.S.Weather Rep. Technical Paper, 48.

(8) Heaps,N.S.,1967: Storm Surges. Oceanogr. Mar. Biol. Ann. Rev. Lon.

(9) Heaps,N.S.,1969: A two-dimensional numerical sea model.
     Phil. Trans. R. Soc. London.

(10) Henning,D.,1961: Computation of storm surge in the Baltic Sea.
     Proc. on Mathematical-Hydrodyn. Methods of Phys.Ocean.Hamb.

(11) Hunt, R.D.,1972: North Sea storm surges. Marine Observer,XLII.

(12) Jankowski,a. and Kowalik,Z.,1978: Wind-driven currents and river
     inflow circulation in the homogeneous Baltic Sea.
     Paper presented at the XI Conference of the Baltic Oceanogr.

(13) Jelesnianski,Ch.P.,1978: Storm surges. Geophysical Predictions,
     Studies in Geophysics, National Academy of Sciences.

(14) Kamenkovitch,V.M.,1978: Wind induced currents in homogeneous
     ocean. Oceanology, Physics of the Ocean, Vol.2.

(15)   Kowalik,Z.,1969: Wind driven circulation in a shallow sea with
          application to the Baltic Sea -I. Acta Geophys.Polonica
          Vol.21, 1.

(16)   Laska,M.,1966: The prediction problem of storm surges in the
          Baltic Sea based on numerical calculations, Arch.hydr.13.

(17)   Laska,M.,1969: Characteristics of storm surges.(in Polish).Techn.
          Gosp. Morska, 3-9.

(18)   Leendertse,J.J.,1967: Aspects of computational model for long
          period water-wave propagation. The Rand Corporation.

(19)   Pore,N.A.,1961: The Storm Surge. Mariners Weather Log. Vol.5 No 5.

(20)   Ramming,H.G.,1970: Investigation of motion processes in shallow
          water areas and estuaries. Symp. on Coastal Geod. Munich.

(21)   Sündermann,J.,1974: Hydrodynamics of Bays and coastal waters.
          Int. Centre of Mechanical Sciences, Udine, Italy.

(22)   Svansson,A.,1959: Some computations of water heights and currents
          in the Baltic. Tellus,2.

(23)   Tomczak, G.,1954: Der Windstau und Sturmflutwarndienst für die
          Deutsche Nordseeküste beim Dt.Hydrogr.Inst. Dt.hydrogr.Z.7.

(24)   Uusitalo,S.,1961: The numerical calculation of wind effect on sea
          level elevation. Proc. on Math.,Hydrod. Methods of Phys.
          Oceanography, Hamburg.

(25)   Welander,P.,1961: Numerical prediction of storm surges. Advances
          in Geophysics, Vol. 8.

# METEOROLOGICAL PROBLEMS ASSOCIATED WITH NUMERICAL STORM SURGE PREDICTION

E. Roeckner

Meteorologisches Institut der Universität Hamburg

## 1. Introduction

There is increasing tendency in most North Sea states to use numerical models of the atmosphere and the sea for the purpose of storm surge prediction. This development was accelerated by three facts, mainly, the availability of high speed computers, the refinement of the models and, finally, the occurrence of severe storm surges during the last years. The results obtained, for example, by Davies and Flather (1977) for April 1973 suggested that the technique should be examined in operational surge prediction. A similar scheme was developed at the University of Hamburg and will now be tested extensively together with the German Weather Service and the German Hydrographic Institute.

## 2.    The models

The hydrodynamic model of the North Sea and the English Channel has a resolution of $\Delta\lambda = 1/3°$ and $\Delta\varphi = 1/5°$ ( $\lambda$ = longitude, $\varphi$ = latitude), corresponding to a mean grid distance of about 20 km.  At the open boundaries the M2-tide is prescribed.  These boundary values are improved by considering external surge effects calculated from a coarse North Atlantic Ocean model without tides (Dolata, 1978).

The atmospheric model covers the northern hemisphere with a grid of $\Delta\lambda = 2.8°$ and $\Delta\varphi = 1.4°$, corresponding to a mean spacing of about 150 km in middle latitudes (Roeckner, 1978).  In the vertical, the total atmosphere is subdivided into 8 constant mass layers.  Physically, the model describes the large scale hydro- and thermodynamic structure of the atmosphere including subgrid-scale processes in parameterized form.  A realistic but slightly smoothed orography is included which not only influences the atmospheric flow dynamically but also indirectly, predominantly in the boundary layer, by an increased surface roughness.

The numerical scheme was developed from the viewpoint of conservation properties in long term integrations, however, the scheme turned out to be suitable also for short range weather forecast experiments if the resolution is chosen sufficiently fine.

3. <u>The method</u>

A first description of our method with some preliminary results was given by Fischer (1978). The main points may be summarized as follows:

(i)    The atmospheric model starts with <u>initial data</u> provided by an objective analysis of observational data by the German Weather Service.    After interpolation to the model's grid the data are subjected to a balancing procedure.

(ii)   The variables <u>forecasted</u> by the model (pressure, wind, temperature and humidity) are stored in 1/2 - hourly intervals in the North Sea - and Atlantic Ocean area, providing the atmospheric input for the sea model.

(iii)  The atmospheric variables are <u>interpolated</u> in space from the 150 km meteorological grid to the 20 km North Sea grid and also in time from 1/2 - hourly values to 3 min intervals for the North Sea model.

(iv)   <u>Surface stress</u> is calculated in the sea model grid from the interpolated atmospheric variables.

(v)    The <u>surface elevations</u> predicted by the sea model are compared with measurements at 16 gauge stations in the North Sea.

## 4. Meteorological problems

From the meteorological point of view, it is extremely difficult
to predict the intensification, the track and the decay of a storm cyclone correctly, for three reasons:

Firstly, the observational network is very coarse over the oceans
where storm cyclones tend to develop. If, however, the structure of
the cyclone is not sufficiently described initially, the model has no
chance of predicting its development correctly.

Secondly, the observed mass- and windfields contain local effects giving rise to unrealistic gravitational oscillations in a large scale
weather prediction model. The necessary removal of these oscillations,
however, changes the information slightly thus increasing the initial
error. Furthermore, the numerical model needs some time to find a
suitable balance of forces acting in it. During this time in which
the overall error evolution is largest, rapid developments are generally slowed down. This time lag effect very often accounts for an
underestimation at the beginning and an overestimation after the observed climax of a storm.

Thirdly, the models contain mathematical and physical shortcomings,
for instance, truncation errors due to the discrete representation
of model variables or the parameterization of subgrid-scale transports of heat, water vapour, and momentum, especially in the atmospheric boundary layer.
The first two points, the initial data and balancing problems are
largely beyond the scope of numerical modellers. The third point,
however, the model deficiencies, are subject for further investigation. The most important aspects with respect to storm surge prediction are the horizontal and vertical resolution and the surface stress
calculation.
The horizontal grid should be in the order of 50 km, which can be
achieved at present only in a nested or limited area model. This relatively high resolution is desirable for two reasons: Firstly, atmospheric models are highly sensitive to truncation errors because of
the dominance of nonlinear effects in atmospheric dynamics and, secondly mesoscale effects like fronts, which significantly affect the
surface wind field of a storm cyclone should be resolved.
Storm cyclone developments are highly sensitive to the vertical structure of the atmosphere. If this structure cannot be described adequately by the model a correct forecast would not be possible. A vertical resolution of about 10 layers for the troposphere seems to be

necessary. However, for storm surge prediction demanding for an accurate simulation of surface layer processes, the atmospheric boundary layer should be resolved by additional 4 or 5 layers resulting in a total number of about 15 layers for a storm cyclone prediction model. A further reduction in mesh size, horizontally or vertically, is probably unnecessary unless the density of the meteorological observing system is increased. Moreover, a model with a grid spacing of 50 km horizontally and 15 layers vertically already exceeds the computer capacity of most weather services. Under the present conditions of observing network such models can only define the upper limit of accuracy attainable in storm surge prediction.

A general problem in all weather prediction models is the parameterization of convective and turbulent subgrid-scale processes. For storm surge prediction, the most important aspect is the vertical structure of the turbulent momentum flux in the boundary layer, which determines the surface wind stress, finally.

The problem of calculating its horizontal distribution is closely related to the vertical resolution of the model. Practically all prediction models, like ours, do not resolve the boundary layer - for economical reasons. The surface stress has to be calculated from the variables in the lowest model level (about 500 m high) by application of some boundary layer theory. These theories, however, are not very well developed and verified by measurements above the surface layer (about 100 m high). Consequently, surface stress calculations from model variables above the surface layer are not very accurate. The aim should be, in storm surge prediction, to place the lowest model level within the surface layer, say 50 m above the surface. By application of well developed surface layer theory it would be very easy and reliable to calculate the surface stress.

Our model, at present, has only 8 layers totally, so it is not possible to resolve the boundary layer. For storm surge prediction we have used two types of methods:

(i) Empirical method

The most popular way of calculating the surface stress used for the sea model is by assuming a linear relationship between the surface wind $V_0$ and the geostrophic wind $V_{go}$ which is either predicted by an atmospheric model or taken from observations, for hindcasts (Hasse, 1974)

$$|V_0| = (0.54 - 0.012\Delta\theta)|V_{go}| + 1.68 - 0.015\,\Delta\theta \qquad (1)$$

The cross isobar angles ranges between $8^\circ$ and $23^\circ$ depending on stability. The surface stress $\vec{\tau}_s$ is calculated from a bulk formula with
a drag coefficient $C_D$ depending on the magnitude of the geostrophic
wind

$$\vec{\tau}_s = \rho_o \, C_D \, \gamma^2 \, |W_o| \, W_o \tag{2}$$

$$C_D = (1.18 + 0.016 \, |W_{3o}|) \cdot 10^{-3} \tag{3}$$

with $\gamma^2 = \begin{cases} 1 & \text{for hindcasts} \\ 1.55 & \text{for forecasts} \end{cases}$

The constant $\gamma^2$ is used at present to correct the the total forecast
error of the meteorological model which results essentially in an underestimation of pressure gradient.

(ii)  Theoretical method

If the atmospheric boundary layer is assumed to be homogeneous
horizontally and stationary, the surface fluxes can be derived from
similarity theory resulting in the so called resistance law.

$$\frac{u_g}{u_*} = \frac{1}{k} \ln \left( \frac{u_*}{f \, z_o} \right) - \frac{A}{k} \tag{4}$$

$$\frac{v_g}{u_*} = - \frac{B}{k} \tag{5}$$

( $f$ = Coriolis parameter, $z_o$ = roughness length, $k$ = Kármán's constant, $u_g$ and $v_g$ are the components of the geostrophic wind  and $u_*$
is the friction velocity).
The empirical functions A, B depending on stability and baroclinici-
ty have to be deduced from measurements or, from a boundary layer
model.  In our case, A and B are derived from an analytical boundary
layer model taking also into account instationary effects derived
from the large scale forcing of the boundary layer (Hecht, 1978).
If applied to storm surge calculation both methods give similar re-
salts, as apparent from figure 1 which shows the hindcasted surge
profile for the case of January 3, 1976.  Both methods largely agree
also if applied to forecasts so that it may be stated that the atmospheric model's main information for the sea model is the geostrophic wind.  The atmospheric model is too coarse vertically to give

additional information about the structure of the boundary layer.
Possibly, a true alternative to both methods would be the explicit
resolution of the boundary layer as mentioned above.

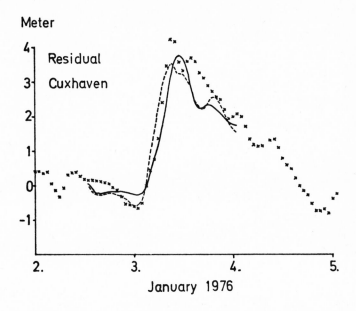

Figure 1    Hindcasted surge profiles for Cuxhaven according to
            resistance law    (———)
            Hasse's relation (-----)
            crosses mark observations

5.    Results

     Some of the problems mentioned above may be illustrated by the
results we have got in testing our procedure described in chapter 3.
for the case of January, 1976.  Figure 2 shows the RMS-error evolution
of the geostrophic wind over the North Sea for two model versions A
and B compared also with the German Weather Service (DWD) forecast.

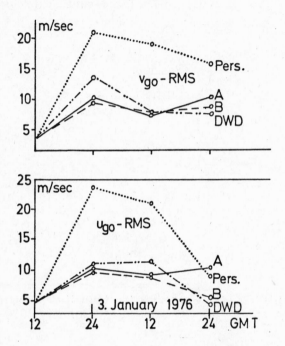

Figure 2    RMS - errors of geostrophic wind in the North Sea area

Model A assumes a smooth sea  everywhere (no land/sea contrast in surface roughness) and does not include condensational processes.  Model
B, on the other hand, contains both aspects.  Both models have the
same resolution of about 150 km horizontally and 8 layers vertically
in contrast to a resolution of 381 km and 6 layers.  The persistence
forecast given for comparison defines the skill of the models. A few
points are noteworthy in figure 2.
(i)    Compared to the total error of about 10 m/sec in geostrophic
       wind, the differences between the models themselves are small,
       generally.
(ii)   The inclusion of land/sea contrast in surface roughness (model
       B) improves the forecast during the decaying phase of the

cyclone after 24 hours of prediction time.

(iii) The relatively coarse DWD-model underestimates the cyclone's development.

(iv) The initial state is not very well defined. The initial uncertainty of about 5 m/sec in geostrophic wind was derived by differences between the objective DWD-analysis used as initial state for the forecast and the re-analyzed surface pressure field using additional data which were not available for the original analysis.

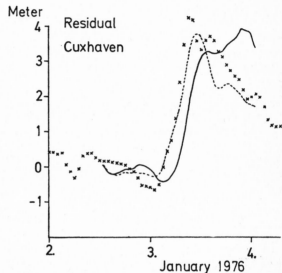

**Figure 3**

Forecast of surge profile for Cuxhaven (———) compared to hindcast (----) and observations (crosses)

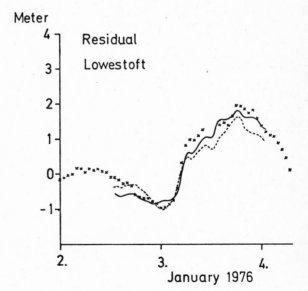

**Figure 4**

The same as Figure 3 for Lowestoft

The geostrophic wind errors in the forecast are largely due to an underestimation of pressure gradients with the consequence of an underestimation of surface elevation if applied to storm surge prediction in a sea model. If, however, a correction factor according to (2) is applied, the predicted surge profiles look reasonable compared with measurements (figures 3 and 4). The best forecast was achieved for Lowestoft and the worst for Cuxhaven.

From the meteorological point of view, the aim for the future should be to improve the models in a way described in chapter 4.

References

Davies,A.M. and R.A.Flather, 1977: Computation of the storm surge of 1 to 6 April 1973 using numerical models of the North West European continental shelf and the North Sea. DHZ 30,5,139-162

Dolata,L.F., 1978: Personal communication

Fischer,G., 1978: Results of a 36-hour storm surge prediction of the North Sea for January 3, 1976 on the basis of numerical models. To appear in DHZ.

Hasse,L., 1974: On the surface to geostrophic wind relationship at sea and the stability dependence of the resistance law. Contr.Atm.Phys.45, 45-58

Hecht,H., 1978: Personal communication

Roeckner,E., 1978: A hemispheric model for short range numerical weather prediction and general circulation studies. Submitted to Contr.Atm.Phys.

A COASTAL OCEAN NUMERICAL MODEL

Alan F. Blumberg and George L. Mellor

Geophysical Fluid Dynamics Program
Princeton University
Princeton, NJ 08540

ABSTRACT

A coastal ocean model which, it is believed, is advanced beyond the current state of
the art has been developed but is only in an early stage of application. Characteristics of the model include:

 * a second moment turbulence closure model capable of accurate prediction of small
   scale turbulent mixing and derivative ocean features such as mixed layer temperature and depth.

 * an algorithm which calculates the external (tidal) mode separately from the internal mode. The external mode, an essentially two-dimensional calculation, requires
   a short integrating time step whereas the costly, three-dimensional, internal mode
   can be executed with a long step. The result is a fully three-dimensional code
   which includes a free surface at no sacrifice in computer cost compared to rigid
   lid models.

 * a "$\sigma$" coordinate system with 20 levels in the vertical independent of depth. Thus,
   the environmentally important continental shelf, shelf bank and slope will be well
   resolved by the model. Furthermore, the model features increased resolution in the
   surface and bottom layers.

 * coding deliberately designed for modern array processing computers. This is essential to three-dimensional ocean simulations requiring long integrations at toler-
   able cost.

# INTRODUCTION

In the last several years, second moment models of small scale turbulence have been developed at Princeton University (Mellor, 1973; Mellor and Durbin, 1975; Mellor and Yamada, 1974) such that mixing or the inhibition of mixing of momentum, temperature and salinity (or any other ocean property) can be predicted with considerable confidence. A number of other investigations have tested the simplest version of the model (Martin, 1976; Martin and Roberts, 1977; Weatherly and Martin, 1978) and it is now a part of the large, weather and climate General Circulation Models at NOAA's Geophysical Fluid Dynamics Laboratory (Miyakoda and Sirutis, 1977).

Incorporating an advanced version of this turbulence model (Mellor and Yamada, 1977), a three dimensional, time dependent, numerical ocean model has been recently constructed which, it is believed, is considerably advanced beyond that which is otherwise currently available. Mean velocity, temperature, salinity, turbulent kinetic energy and turbulent macroscale are prognostic variables. Free surface elevation is also calculated prognostically with no sacrifice in computational time. The model incorporates a "σ" coordinate system such that the number of grid points in the vertical is independent of depth. Furthermore, the spacing in this transformed coordinate system is also variable so that, for example, one may stipulate finer resolution near the surface and bottom layers resulting in an algorithm which will be very economical on modern array processing computers.

The model responds to tidal forcing, surface wind stress, heat flux, salt "flux" (i.e., evaporation minus precipitation), estuarine outflow and to the specification of temperature, salinity and sea surface elevation at open inflow boundaries.

At this writing, the model has just become operational. Some coastal ocean simulations are presented in this paper but the real effort of comparing data and calculation lies ahead.

DESCRIPTION OF THE NUMERICAL MODEL

Model Physics

The equations of motion which are solved by the model are:

$$\frac{\partial U}{\partial t} + \underline{V} \cdot \underline{\nabla}U - fV = -\frac{1}{\rho_o}\frac{\partial P}{\partial x} + \frac{\partial}{\partial z}\left[K_M \frac{\partial U}{\partial z}\right] + F_x \tag{1a}$$

$$\frac{\partial V}{\partial t} + \underline{V} \cdot \underline{\nabla}V + fU = -\frac{1}{\rho_o}\frac{\partial P}{\partial y} + \frac{\partial}{\partial z}\left[K_M \frac{\partial V}{\partial z}\right] + F_y \tag{1b}$$

$$0 = -\frac{\partial P}{\partial z} + \rho g \tag{2}$$

$$\frac{\partial U}{\partial x} + \frac{\partial V}{\partial y} + \frac{\partial W}{\partial z} = 0 \tag{3}$$

$$\frac{\partial T}{\partial t} + \underline{V} \cdot \underline{\nabla}T = \frac{\partial}{\partial z}\left[K_H \frac{\partial T}{\partial z}\right] + F_T \tag{4}$$

$$\frac{\partial S}{\partial t} + \underline{V} \cdot \underline{\nabla}S = \frac{\partial}{\partial z}\left[K_H \frac{\partial S}{\partial z}\right] + F_S \tag{5}$$

where U, V, T, S are the mean velocity components, temperature and salinity and we define $\underline{V} \cdot \underline{\nabla}( ) \equiv U\partial( )/\partial x + V\partial( )/\partial y + W\partial( )/\partial z$. The turbulence field is characterized by

$$\frac{\partial q^2}{\partial t} + \underline{V} \cdot \underline{\nabla}q^2 = \frac{\partial}{\partial z}\left[K_q \frac{\partial q^2}{\partial z}\right] + 2 (P_s + P_b) - 2 \varepsilon + F_q \tag{6}$$

$$\frac{\partial}{\partial t} (q^2\ell) + \underline{V} \cdot \underline{\nabla}(q^2\ell) = \frac{\partial}{\partial z}\left[K_q \frac{\partial}{\partial z} (q^2\ell) \right] + \ell E_1 (P_s + E_3 P_b) - \ell \varepsilon \widetilde{W} + F_\ell \tag{7}$$

where $q^2/2$ is the turbulent kinetic energy; $\ell$ is a turbulent macroscale; $P_s$ and $P_b$ are turbulent shear and buoyancy production; $\varepsilon$ is dissipation and $\widetilde{W}$ is a wall proximity function. The problem is primarily closed by expressions for $K_M$, $K_H$, and $K_q$

which are function of $\partial U/\partial z$, $\partial V/\partial z$, $\rho_o^{-1}$ $g\partial\rho/\partial z$, $\ell$ and q. These are analytically derived relations emanating from closure hypotheses described and implemented by Mellor (1973), Mellor and Yamada (1974), Yamada and Mellor (1975) and most recently by Mellor and Yamada (1977). Appendix A contains most of the details. Empirical constants in these algebraic relations are derived from neutral data but the result has been shown to predict the stabilizing or destabilizing effects of density strat-ification. The density, $\rho$, is of course related to temperature and salinity through an equation of state for sea water.

The terms, $F_x$, $F_y$, $F_T$, $F_S$, $F_q$ and $F_\ell$ represent horizontal diffusion which are usually required by models to damp small scale numerical computational modes. Oftentimes, the required horizontal diffusivities give rise to excessive smoothing of real oceanog-raphic features. The problem is, of course, ameliorated by decreasing horizontal grid size. In our case, we believe that relatively fine *vertical* resolution results in a reduced need for *horizontal* diffusion; i.e., horizontal advection followed by vertical mixing effectively acts as horizontal diffusion in a real physical sense.

Boundary Conditions

The boundary conditions at the free surface, $z = \eta(x,y)$, are:

$$K_M \left( \frac{\partial U}{\partial z}, \frac{\partial V}{\partial z} \right) \sim (\tau_{ox}, \tau_{oy}) \text{ as } z \to \eta \qquad (8a,b)$$

$$K_H \left( \frac{\partial T}{\partial z}, \frac{\partial S}{\partial z} \right) \sim (\dot{H}, \dot{S}) \quad \text{ as } z \to \eta \qquad (9a,b)$$

$$q^2 = B_1 u_\tau^2, \quad z = \eta \qquad (10)$$

$$q^2\ell = 0, \quad z = \eta \qquad (11)$$

$$W = U \frac{\partial \eta}{\partial x} + V \frac{\partial \eta}{\partial y} + \frac{\partial \eta}{\partial t}, \quad z = \eta \qquad (12)$$

where $(\tau_{ox}, \tau_{oy})$ is the surface wind stress vector, $\dot{H}$ is the net ocean heat flux and $\dot{S} = S(0)[\dot{E}-\dot{P}]/\rho_o$ where $(\dot{E}-\dot{P})$ is the net evaporation-precipitation fresh water surface flux rate. In equation (10), $u_\tau^2 = |\tau_o|$ and $B_1$ is one of the empirical constants in the turbulent closure relations.

At the bottom, $z = - H(x,y)$, boundary conditions for T, S, $q^2$ and $q^2\ell$ are similar to

(9a,b), (10) and (11) where, however, $\dot{H} = \dot{S} = 0$. In place of (12) we have $W = -U\partial H/\partial x - V\partial H/\partial y$ where $H(x,y)$ is the bottom topography. Bottom boundary conditions for U and V are supplied by matching the solution to the logarithmic law of the wall which requires a bottom roughness parameter. In deep water, bottom boundary layers may be unimportant, but may assume some importance on the shelf. However, some recent work by Armi (1978) indicates that bottom boundary layers are important for the long time scale development of the thermocline. The hypothesis is that bottom boundaries on, say the continental slopes, mix adjacent vertical layers of water which are then advected into the interior. According to the hypothesis this effect may be more important than small vertical mixing attributable to internal gravity wave breaking, at least, in deeper water well below the mixed layer. Our model, in principle, can account for this behavior.

In the Middle Atlantic Bight simulation discussed later, open boundary conditions require temperature and salinity. Geostrophically derived, vertical gradients of horizontal velocity may be calculated but then either total transport or sea surface elevation is also required.

## Numerical Scheme

To achieve computational economy the program is divided into external and internal mode subprograms. The first, call it the XYt subprogram, computes the vertically averaged velocity and the surface elevation fields with a short time increment ($\approx 30$ sec.) imposed by the shallow water wave speed, CFL criterion, The second, call it the XYZT subprogram, computes the full three-dimensional velocity, temperature and salinity fields with a much longer time increment ($\approx 40$ minutes). The XYZT subprogram incorporates the second moment turbulent closure model. It supplies computed bottom friction and vertical integrals of density and vertical variances of horizontal velocity to the XYt subprogram where they behave as lateral friction-like terms in the vertically averaged horizontal equations of motion. (These terms must be parameterized by horizontal eddy viscosities in models which do not adequately resolve vertical structure.) In turn, the XYt subprogram supplies sea surface evaluation to the XYZT subprogram. This may sound complicated, but in the final analysis, the full, three-dimensional field equations are solved with a free surface boundary condition at no additional cost in computer time as compared to rigid lid models (Bryan and Cox, 1968).

The time differencing is the conventional leap frog technique. However, the scheme is quasi-implicit in that vertical diffusion is evaluated at the forward time level. Thus, small vertical spacing is permissible near the surface without need to reduce the time increment or restrict the magnitude of the mixing coefficients.

The vertical coordinate is scaled such that $\sigma = (z-\eta)/(H+\eta)$ and all equations are transformed to $x,y,\sigma,t$. Currently, we use 20 vertical levels with increasingly fine resolution near the surface and bottom so that surface and bottom mixed layers are resolved. The resolution in physical space increases shoreward as H decreases.

## Present Status of the Model

In the process of developing the model some intrinsically interesting exploratory calculations have been made. The initial numerical experiments involve the 2-D, XYt mode (all longshore gradients are neglected) to simulate the effects of coastal up-welling and downwelling. Figures 1, 2 and 3 illustrate the results of an impulsively imposed alongshore wind stress. Three cases are considered: Figure 1 is a homogeneous, upwelling event, Figure 2 a density stratified, upwelling event, and Figure 3 is a density stratified, downwelling event. The role of stratification is confining mixing to surface and bottom layers is readily apparent. In Figure 2 one will observe the formation of a near shore (x = 2km) baroclinic jet.

The numerical code has also been exercised in the external (tidal) mode. An application of this mode to the Chesapeake Bay (Blumberg, 1977) showed considerable success. The 2-D tidal mode also has been applied to the Middle Atlantic Bight (1/4° horizontal resolution). Figure 4 illustrates the dynamic response of a "barotropic" MAB to various surface elevation boundary conditions imposed along the open portions of the domain.

The fully three-dimensional code has only very recently become operational with two time steps (recall that the external, vertically averaged mode requires a short time step, whereas the fully three-dimensional calculations can be executed with an economically long time step) after a long debugging period. Figure 5 is the result of a calculation of the Middle Atlantic Bight circulation with manufactured temperature and salinity distributions for initialization and for open boundary conditions. The normal component of velocity along the open boundary is specified by geostrophic balance with a level of no motion at the bottom. Also, for this calculation the surface wind stress and fluxes are zero. A transect east of Cape Hatteras is shown in Figure 6; contours of north/south velocity are drawn in this diagram.

Numerical experiments are now being conducted using the climatological temperature and salinity distributions described by Blumberg, Mellor and Levitus (1978) as initial conditions and at the open boundaries. Preliminary prognostic simulations (temperature, salinity and therefore density are simulated) show a broad, slow Gulf Stream together with a southward flow along the coast. The velocity distributions spin-up in about 5 days; however, the temperature and salinity fields evolve more slowly.

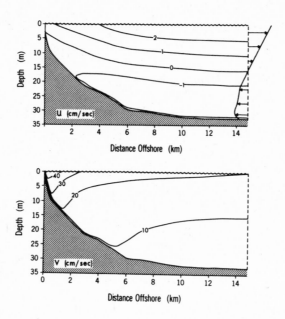

Figure 1. A homogeneous upwelling event induced by an alongshore wind stress of 2.0 dynes/cm$^2$ directed into the plane of the paper. The wind stress has been imposed for six hours. The onshore (U negative) and offshore (U positive) isotachs are depicted in the upper portion of the figure, while the alongshore (V positive into the plane of the paper and V negative out of the plane of the paper) isotachs are depicted in the lower portion.

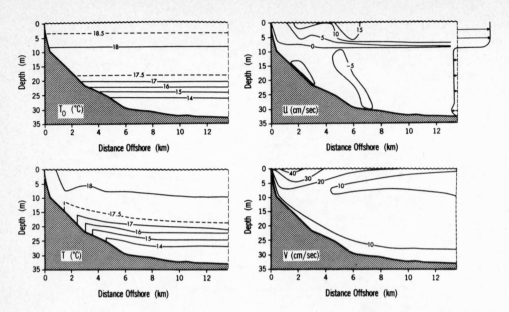

Figure 2. A stratified upwelling event induced by an alongshore wind stress of 1.0 dyne/cm directed into the plane of the paper. This wind stress has been imposed for twelve hours. The direction of the isotachs is the same as in figure 1. The initial temperature distribution is denoted as $T_0$.

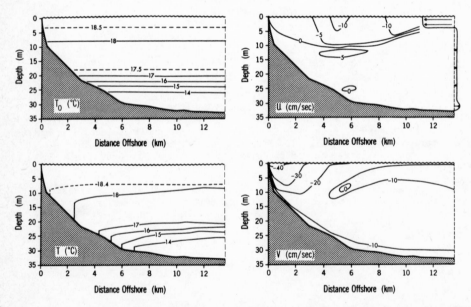

Figure 3. A stratified downwelling event induced by an alongshore wind stress of 1.0 dyne/$cm^2$ directed out of the plane of the paper. The wind stress has been imposed for twelve hours. The direction of the isotachs is the same as in figure 1. The initial temperature distribution is denoted as $T_0$.

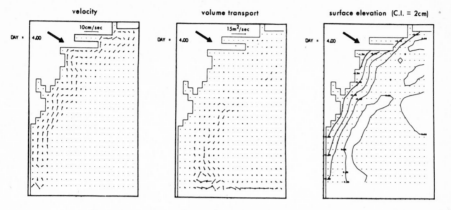

Case A: Zero surface elevation along open boundary

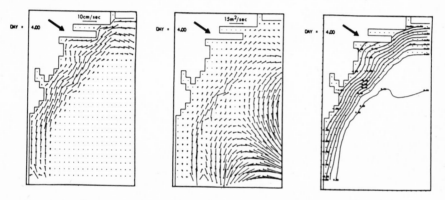

Case B: Modified radiation condition at open boundary

Figure 4. A comparison of the dynamic response of the Middle Atlantic Bight after four days to various surface elevation boundary conditions. The heavy arrow indicates the direction of the 1.2 dyne/cm$^2$ wind stress imposed at Day=0.

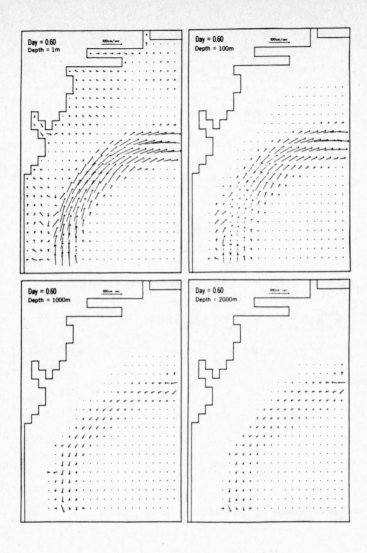

Figure 5. Density driven circulation patterns in the MAB at various depths for a manufactured temperature and salinity distribution.

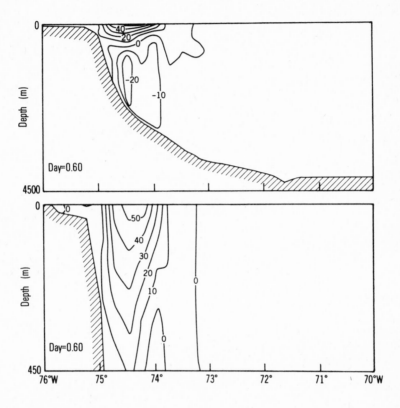

Figure 6. Contours of calculated North/South velocity (isotachs in cm/sec) on Latitude 36°N. The lower figure is a detail of the upper 450m.

CONCLUSION

The construction and implementation of a fully three-dimensional numerical model capable of predicting the dynamics and thermodynamics of coastal ocean regions is presented. Genuine simulations for comparison with real data have yet to be initiated and will, in fact, be the major goal of future research.

ACKNOWLEDGMENTS

This work is a result of research sponsored by NOAA Office of Sea Grant, Department of Commerce, under Grant # 04-6-158-4476, and NOAA/Princeton University, Visiting Scientist Grant # 04-7-022-4417.

## APPENDIX A

Since the paper by Mellor and Yamada (1974), a few modifications have been made to the boundary layer model.

First, the "Level 3" model was further simplified into a "Level 2 1/2" model by neglect of the material and diffusive derivatives for scalar (temperature, salinity, density, etc.) variances. The loss in predictive accuracy is not expected to be important (Yamada, 1977).

Second, as discussed by Mellor and Yamada (1977), the empirical constants cited below have been changed slightly from the original values after a critical reexamination of the data upon which they are based. The overall effect of these changes should be quite small indeed.

A third modification is incorporated here and results from trials of a suggestion by Rodi (1972). Consider the model equations for $\overline{u_i u_j}$, $\overline{u_i \rho'}$ and $\overline{\rho'^2}$:

$$\mathcal{L}_1 \left( \overline{u_i u_j} \right) = - \overline{u_k u_i} \frac{\partial U_j}{\partial x_k} - \overline{u_k u_j} \frac{\partial U_i}{\partial x_k} - \frac{2}{3} \frac{q^3}{\Lambda_1} \delta_{ij}$$

$$+ g_j \overline{u_i \rho'} + g_i \overline{u_j \rho'} \tag{A1}$$

$$- \frac{q}{3\ell_1} \left( \overline{u_i u_j} - \frac{\delta_{ij}}{3} q^2 \right) + C_1 q^2 \left( \frac{\partial U_i}{\partial x_j} + \frac{\partial U_j}{\partial x_i} \right)$$

$$\mathcal{L}_2 \left( \overline{u_j \rho'} \right) = - \overline{u_k u_j} \frac{\partial \overline{\rho}}{\partial x_k} - \overline{\rho' u_k} \frac{\partial U_j}{\partial x_k} - g_j \overline{\rho'^2} + \frac{q}{3\ell_2} \overline{u_j \rho'} \tag{A2}$$

$$0 = \overline{u_k \rho'} \frac{\partial \overline{\rho}}{\partial x_k} - \frac{q}{3\ell_2} \overline{\rho'^2} \tag{A3}$$

the operators $\mathcal{L}_1(\ )$ and $\mathcal{L}_2(\ )$ represent the material and diffusive derivative terms. The corresponding terms in (A3) have been neglected in accordance with our previous comments.

Now (A1) upon contraction yields the turbulent energy equation,

$$\mathcal{L}_3 \left( q^2 \right) = 2 (P_s + P_b - \varepsilon) \tag{A4}$$

where, if $g_i = (0, 0, -g)$, production and dissipation are defined according to

$$P_s \equiv - \overline{u_i u_j} \frac{\partial U_i}{\partial x_j}, \quad P_b = - g\overline{w\rho'}, \quad \varepsilon \equiv \frac{q^3}{\Lambda_1} \qquad (A5a,b,c)$$

and $\mathcal{L}_3(\ )$ is very nearly but not exactly identical to $\mathcal{L}_1(\ )$. Rodi's suggestion is to replace $\mathcal{L}_1(\overline{u_i u_j})$ in (A1) with $(\overline{u_i u_j}/q^2)\mathcal{L}_3(q^2)$, where $\mathcal{L}_3(q^2)$ is obtained in (A4). This could have been incorporated into the paper of Mellor and Yamada (1974) as an alternative and apparently equally consistent step by replacing $\delta_{ij}/3$ by $\overline{u_i u_j}/q^2$ at one point in their analysis.[1] Note that if we define $a_{ij}$ such that $\overline{u_i u_j} = (\delta_{ij}/3 + a_{ij})q^2$ then $a_{ij} \to 0$ defines an isotropic limit. Thus, incorporation of Rodi's idea introduces a higher order term into the Mellor-Yamada analysis. This is not incorrect but is somewhat arbitrary since other terms, presumably of the same order, have been neglected. The choice has been made here on the basis that the resulting algorithm survives numerical trauma better than its predecessor and in the hope that the resulting approximation is closer to the full (level 4) equation set.

A similar step is to replace $\mathcal{L}_2(\overline{u_i \rho'})$ with $(\overline{u_i \rho'}/2\ q^2)\ \mathcal{L}_3(q^2)$ since a term like $\mathcal{L}(\overline{\rho'^2})$ on the left of (A3) has already been neglected.

The result of these substitutions is that (A1) and (A2) may be rewritten as

$$2\frac{\overline{u_i u_j}}{q^2}(P_s + P_b - \varepsilon) = - \overline{u_k u_i}\frac{\partial U_j}{\partial x_k} - \overline{u_k u_j}\frac{\partial U_i}{\partial x_k} - \frac{2}{3}\delta_{ij}$$

$$+ g_i\,\overline{u_j \rho'} + g_j\,\overline{u_i \rho'} \qquad (A6)$$

$$- \frac{q}{3\ell_1}(\overline{u_i u_j} - \frac{\delta_{ij}}{3}q^2) + \tilde{C}_1 q^2 (\frac{\partial U_i}{\partial x_j} + \frac{\partial U_j}{\partial x_i})$$

$$\frac{\overline{u_i \rho'}}{q}(P_s + P_b - \varepsilon) = - \overline{u_j u_k}\frac{\partial \overline{\rho}}{\partial x_k} - \overline{\rho' u_k}\frac{\partial U_i}{\partial x_k} - \frac{q}{3\ell_2}\overline{u_i \rho'}$$

$$\qquad (A7)$$

$$- g_j\frac{\Lambda_2}{q}\overline{u_k \rho'}\frac{\partial \overline{\rho}}{\partial x_k}$$

where $\overline{\rho'^2}$ in (A2) has been replaced using (A3).

Now, if one makes the boundary layer approximation in (A6) and (A7), all components

---

[1] Thus on page 1793 of the Mellor-Yamada paper, in the first sentence below equation (8), substitute $\overline{u_i u_j}/q^2$ in place of $\delta_{ij}/3$.

of the tensor, $\partial U_j/\partial x_k$, may be neglected except for $\partial U/\partial z$ and $\partial V/\partial z$. Then, if we again set $g_i = (0, 0, -g)$, and define $-(\overline{uw}, \overline{vw}) = K_M(\partial U/\partial z, \partial V/\partial z)$ and $-\rho'w \equiv K_H \partial\rho/\partial z$, and furthermore define

$$K_m \equiv \ell q\, S_M, \qquad K_H \equiv \ell q\, S_H \tag{A8a,b}$$

and

$$\phi_1 \equiv 1 + \frac{3\ell_1}{q^3}(P_s + P_b - \varepsilon) = 1 + \frac{3\ell_1}{\Lambda_1}\left(\frac{P_s + P_b}{\varepsilon} - 1\right) \tag{A9a}$$

$$\phi_2 \equiv 1 + \frac{3\ell_2}{q^3}(P_s + P_b - \varepsilon) = 1 + \frac{3\ell_2}{\Lambda_1}\left(\frac{P_s + P_b}{\varepsilon} - 1\right) \tag{A9b}$$

$$C_1 \equiv \frac{\widetilde{C}_1}{\phi_1} \tag{A9c}$$

$$G_H \equiv \frac{\ell^2}{q^2}\frac{g}{\rho_o}\frac{\partial\overline{\rho}}{\partial z} \tag{A9d}$$

then

$$S_M'\left[\phi_1 - \frac{9A_1A_2G_H}{\phi_2}\right] - S_H\left[\frac{18A_1^2G_H}{\phi_1} + \frac{9A_1A_2G_H}{\phi_2}\right]$$

$$\tag{A10}$$

$$= A_1\left[\frac{1 - 6A_1/B_1 - 3C_1}{\phi_1}\right]$$

$$S_H\left[\phi_1 - 3A_2\,B_2\,G_H - 18A_1\,A_2\,G_H\right] = A_2\left[\frac{1 - 6A_1/B_1}{\phi_1}\right] \tag{A11}$$

Equation (A9c) where $C_1$ = constant is another numerically motivated modification and has no physical basis. Generally $\widetilde{C}_1 \simeq C_1$ except in start up situations or below the mixed layer where $q^2$ and other variables are dominated by round-off error.

It may be shown that the correlation equations for temperature and salinity are identical to those for density so that

$$-\,\overline{wT'} = K_H\,\partial T/\partial z \quad \text{and} \quad -\,\overline{wS'} = K_H\,\partial S/\partial z.$$

Thus the algebraic forms of (A10) and (A11) have been altered somewhat from the original Level 3 model of Mellor and Yamada (1974, 1977). For $P_s + P_b = \varepsilon$, the model collapses to the same Level 2 (e.g., as in Mellor and Durbin, 1975) model as before. In fact, the present version conforms very nearly to the previous version when $0 < (P_s + P_b)/\varepsilon < 2$ but is numerically more rugged and more readily accommodates start-up shocks, for example.

A necessary assumption is that all lengths are proportional to each other. Thus, we let

$$(\ell_1, \ell_2, \Lambda_1, \Lambda_2) = (A_1, A_2, B_1, B_2)\ell \tag{A12}$$

By appealing to simple laboratory data (Mellor and Yamada, 1977) all of the empirical constants were assigned the values;

$$(A_1, A_2, B_1, B_2, C_1) = (.92, .74, 16.6, 10.1, 0.08) \tag{A13}$$

There remain unknowns in equations (6) and (7). First, in analogy to (A8a,b) we define

$$K_q \equiv \ell q \, S_q \tag{A14}$$

Since we have found that $S_M$ and $S_H$ are dependent on stability, one would suppose this to be the case with $S_q$, although a determination of $S_q$ similar to that of $S_M$ and $S_H$ would require an appeal to equations for triple correlations which would require additional modelling assumptions and constants. Thus we have variously tried $S_q$ = constant = 0.20 (determined from neutral boundary layer and channel flow data) and the stability dependent $S_q = 0.20(S_M/0.392)$ where the value $S_M = 0.392$ corresponds to neutral flow and where production is balanced by dissipation. However, results are not overly sensitive to this choice. In this paper, we have used the second prescription for $S_q$.

In equation (7), $\widetilde{W}$ is a "wall proximity" function defined as

$$\widetilde{W} = 1 + E_2 \left(\frac{\ell}{L}\right)^2 \tag{A15}$$

L has a more general definition but for the ocean problem $(\kappa L)^{-1} = (\eta - z)^{-1} + (H+z)^{-1}$. Near surfaces it may be shown that both $\ell$ and L are proportional to distance from the surface ($\kappa = 0.4$ is the constant of proportionality) and therefore $\widetilde{W} = 1 + E_2$; far from surfaces $\widetilde{W} = 1$.

Finally, the following constants have been determined from simple laboratory boundary layer and channel flow data (Mellor and Yamada, 1977) such that

$$(E_1, E_2, E_3) = (1.8, 1.33, 1.0) \tag{A16}$$

$E_3 = 1.0$ is a default value awaiting a comparison of a data set and calculation that would discrimate an alternative value.

REFERENCES

Armi, L., 1978: Some evidence for boundary mixing in the deep ocean. *J. Geoph. Res.*, 83, 1971-1979.

Blumberg, A., 1977: Numerical tidal model of Chesapeake Bay. *J. of the Hydraulic Division.*, ASCE, 103, (HY1), 1-10.

Blumberg, A.F., G.L. Mellor, and S. Levitus., 1977: The Middle Atlantic Bight: A climatological atlas of oceanographic properties. Princeton University Sea Grant Report No. NJ/P-SG-01-8-77.

Bryan, K., and M.D. Cox., 1968: A nonlinear model of an ocean driven by wind and differential heating. Parts I and II. *J. Atmos. Sci.*, 25, 945-978.

Martin, P.J. 1976: A comparison of three diffusion models of the upper mixed layer of the ocean. NRL-GFD/OTEC 4-76, 1-56.

Martin, P.J. and G.O. Roberts. 1977: An estimate of the impact of OTEC operation on the vertical distributions of heat in the Gulf of Mexico. Proceeding, Fourth Annual Conference on Ocean Thermal Energy Conversion, New Orleans.

Mellor, G.L., 1973: Analytic prediction of the properties of stratified planetary surface layers. *J. Atmos. Sci.*, 30, 1061-1069.

Mellor, G.L. and T. Yamada., 1974: A hierarchy of turbulence closure models for planetary boundary layers. *J. Atmos. Sci.*, 31, 1791-1806.

Mellor, G.L. and P.A. Durbin, 1975: The structure and dynamics of the ocean surface mixed layer., *J. Phys. Oceanogr.*, 5, 718-728.

Mellor, G.L. and T. Yamada, 1977: A turbulence model applied to geophysical fluid problems. Proceedings: Symposium on Turbulent Shear Flows, Pennsylvania State University, University Park, Pennsylvania, April 1977, 1-14.

Miyakoda, K. and J. Sirutis, 1977: Comparative integrations of global models with various parameterized processes of subgrid-scale vertical transports: Description of the parameterizations. *Beitrage zur Physik der Atmosphare.*, 50, 445-487.

Rodi, W., 1972: The prediction of free turbulent boundary layers by use of a 2-equation model of turbulence. Ph.D. thesis, University of London.

Weatherly, G. and P.J. Martin., 1978: On the structure and dynamics of the oceanic bottom boundary layer. *J. Phys. Oceanogr.*, 8, 557-570.

Yamada, T. and G.L. Mellor., 1975: A simulation of the Wangara atmospheric boundary layer data., *J. Atmos. Sci.*, 32, 2309-2329.

Yamada, T., 1977: A numerical experiment on pollutant dispersion in a horizontally-homogeneous atmospheric boundary layer., *Atmosphere Environment.*, 11, 1015-1024.

# MODELLING AND VERIFICATION OF CIRCULATION

## IN AN ARCTIC BARRIER ISLAND LAGOON SYSTEM -

## AN ECOSYSTEM PROCESS STUDY

J. B. Matthews
Geophysical Institute
University of Alaska
Fairbanks
Alaska USA

## Introduction

The topic of this conference is "Mathematical Modelling of Estuarine Physics". Later in the program we shall hear details of the first operational numerical model for storm surge prediction. The modelling of estuarine physics has thus become a daily routine procedure.

Since this is the first paper I shall describe some recent work which may be the direction of future work. In this application estuarine numerical models have been used but are only part of a larger ecosystem model. I have chosen to speak on this topic because I feel that we are going to be asked to contribute to these types of studies in the future. The output from our numerical models will not simply be used to confirm storm surge levels or tidal currents but will be used as predicted environmental data upon which to base other models of ecosystem processes. Philosophically this is an important stage in the development of our field. It means that we lose, as physicists and modellers, some of our independence. However we move into an exciting era of cooperation with other scientists and engineers and become part of the decision-making process for development of estuarine environments.

In 1975 the United States of America began work on a large project to evaluate potential impact on the environment from oil development on the outer continental shelf. Research plans were formulated and requests for proposals sent out. The whole program was controlled by the Bureau of Land Management (BLM) which subcontracted with the National Oceanic and Atmospheric Administration (NOAA) to handle the oceanographic research program. We shall be concerned with this NOAA Outer Continental Shelf Environmental Assessment Program (OCSEAP). Already you can see layers of bureaucracy developing long before we get to the working scientist. The overall OCSEAP program covers all the continental shelves of the United States. The subject of my talk is part of the Alaskan Continental Shelf Program and is managed by the Arctic Project Office of the Alaska program. The project I shall be describing is the brainchild of one man who was part of the NOAA OCSEAP Alaska planning team.

## The need for a new approach

Dr. Herbert Curl Jr. had been a Professor of Marine Biology in the Department of Oceanography in Oregon State University for many years before he became a government employee. When he found himself planning research to assess the impact of oil development in the region off Prudhoe Bay, he was not satisfied that the traditional approach along disciplinary lines would produce research results in a timely manner for use by decision-makers. Part of the reason for this was that the U.S. arctic coastal region was almost unexplored oceanographically. However, there was not time to study every aspect of coastal ecology in the time frame of proposed research. Normally, research contracts are let along disciplinary lines. They run for a few years, then the results of physical, biological, chemical and geological research are reviewed and gaps for further research identified. For energy-hungry USA this approach in 1975 gave insufficient time for a full program of research and evaluation to be undertaken before the offshore tracts were due to be offered for leasing.

Dr. Curl decided to attempt to put together a project which was aimed at iden- tifying those ecosystem processes likely to be impacted by oil development and then to put together an interdisciplinary research program addressing only those aspects expected to be impacted. A major change in this approach was that the research workers would be required to hold frequent work-shops at which new information would be assessed and data gaps identified. The direction of the research efforts could at those times be moved into the most productive lines. It is normal to propose some line of research, be granted funds to accomplish this and then to proceed to completion. This new approach required that research be proposed and funded, and then changed subsequently according to the demands of scientists from other disciplines.

Clearly this ecosystem approach posed problems for both the funding agency and the research workers. Most scientists and beaurocrats predicted its unworkability. Oceanographic work had in its first 100 years generally been performed on disciplinary lines. Researchers are attached to disciplinary departments; their performance is judged on disciplinary lines. It would not be easy to persuade a physical oceanographer to give up a promising line of research because a fisheries expert needs some crude estimates of currents or temperatures.

In short there was no precedent in the USA for this kind of program. Only subsequently have I learned that such an approach has been successfully used in the Netherlands for the Deltaworks project. This latter program probably succeeded because the project employed people specifically to work on the interdisciplinary ecosystem progress study. NOAA OCSEAP was attempting to draw scientists from university, industry and government sources and put together an interdisciplinary research team. Perhaps the key again was financial. The old expression "he who pays the piper calls the tune" applied.

The project has been running since 1976 and it is now generally accepted to be a successful program. It was of course not without major problems. However these have been overcome without changes to the overall research direction. The scien- tific results have already been used in the decision-making process.

Ecosystem process analysis

The basis of the project is ecosystem process analysis. This is defined as a procedure for investigating ecological support mechanisms and inter-dependencies of selected organisms as a primary means of evaluating the vulnerability of these or- ganisms to man-made impacts.

In order to determine the processes likely to be impacted by offshore oil development, it was necessary to know how exploration and development would be con- ducted. It is no easy task for government and university scientists to find out in advance what industry is likely to do in a lease area. Any activity in the region would depend on the oil reservoirs discovered during exploratory drilling, after the lease had been let. Information on oil prospects before a lease sale are closely guarded secrets. We were left with the option of making educated guesses at probable impacts. Later in the research program, we were fortunate to have industry representatives go over our scenarios and suggest changes to help make them more realistic.

Our research suggested that industrial activity would have impacts from several activities. The use of gravel pads 4m thick for roads and drilling sites would take large quantities of gravel. The gravel is essential to serve as an insulator against permafrost. It could come from rivers, beaches or from the sea floor. Its use may take away fish habitat and destroy bird-nesting sites. Drilling pads may be constructed on islands if they are near the oil reservoirs. Otherwise artificial drilling pads would be constructed. For production and development, artificial islands would probably be built of stabilized gravel dredged from the sea bed. For development alone, they could be temporary gravel islands or artificial ice islands. Pipelines and causeways connecting drilling platforms to shore may influence circulation. Drilling muds from offshore platforms increase turbidity locally and may put into the water column, materials toxic to the biota. Last but not least, oil spills could occur at any time during both the exploratory and production phases.

It was believed that the most critical impacts would occur during the open water season when birds and fish are present in abundance. Marine and terrestrial mammals were determined to be relatively unimportant at the outset. The most critical time of all was determined to be the one week in July when large flocks of eider and other birds sit on the lagoon waters, moulting and flightless. There is also a native population which depends on these birds and also on migatory fish for subsistence. The impacts which interfered with these species of birds and fish and their life cycles should be studied by system modelling. Impacts on terrestrial mammals, arctic fox, polar bear and caribou were expected to be minimal. Polar bear denning areas occur to the east of the proposed lease area and probably would not be impacted by oil development. Marine mammals - walrus, seal and whale - generally stay in deeper water and near open leads and the ice edge in summer. Effects on marine mammals from developments in the shallow waters of the lease area were also expected to be negligible.

Ecosystem model

System modelling techniques were used to enable the rapid assimilation of rate and quantity information. The ecosystem model provides numerical estimates of processes, and ties together the research information provided by different disciplines. The model is not a rigorously defined mathematical model of the type we are discussing at this meeting. It is rather a rough assemblage of equations describing the quantities and rates of change of ecosystem components. A numerical model of the physical processes is one of the inputs to the ecosystem model. The ecosystem model is used to determine which processes are important and which need verification and further measurement.

The ecosystem modelling techniques used were developed by Dr. Carl Walters of the Institute of Ecology, University of British Columbia. At the outset, a matrix was drawn up along disciplinary lines. Each discipline, for example, physical oceanographers, meteorologists, ornithologists, benthic invertebrate specialists, sedimentologists, microbiologists and fishery experts, was asked what information it could provide and what information it would need from other disciplines to address the effects of impacts identified earlier. In this way information sources

and gaps were determined.  Generally physical scientists were providers of information rather than receivers.


During the initial workshop it became apparent that  physical  and  biological
processes occur at greatly differing temporal and spatial scales.  Numerical models
of hydrodynamic processes can provide minute-by-minute current or  sea  level  data
over  the  entire  lease  area on,  for  example,  a  1  km grid scale. Fisheries
biologists may have data at daily intervals for only two places along a shore-line.
It  was  necessary  for the physical scientists to provide cruder, averaged data to
satisfy the needs of the other disciplines.

## Study area

The part of the lease area to which this analysis was  applied  is  shown  in
figure 1.   The full lease area extends further to the east for a total distance of
about 135km and is offshore of one of the largest oil fields in  the  USA,  Prudhoe
Bay.   The  potential  for  oil  development  is  obvious.  Simpson Lagoon was chosen
because it is a typical ecosystem of the Beaufort Sea coast, is comparatively undisturbed,  is  reasonably  accessible  and is likely to be impacted by continental
shelf development.  Three research objectives were specified: 1) to  identify  and
analyse  components  and  processes which significantly contribute to the structure
and productivity of the nearshore ecosystem, 2) to evolve  mechanisms  whereby  the
above  components  and processes can be evaluated for their reactions to man-caused
change and 3) to determine the feasibility of detecting  and  quantifying  temporal
change  in  ecosystem  components  and  processes found to be important.  Important
ecosystem components were measured directly.  Measurements  of  processes  are  not
generally direct but appear as rates and amounts of components.


Simpson Lagoon is typical of the Alaskan Beaufort Sea coast.  It has a line of
low-profile  barrier islands 5-10km offshore.  The mainland shoreline is a low tun-
dra plateau dotted with numerous lakes.  The cliffs are only 1-3m  above  mean  sea
level  and coastal erosion averages about 1.4m/year.  The water depth in the lagoon
averages about 2m.  It is frozen from September to late June each year.  Ice thickness  is  about 2m in late spring and only deeper than average channels remain open
year-round.  Offshore of the barrier islands, water depths increase  quite  rapidly
to 30m.  During the open water season the lagoons are breeding and moulting grounds
for many species of birds as well as serving as a refuge and  migratory  route  for
anadramous and pelagic fish.


Prevailing winds are generally from the ENE over the region  and  it  was  anticipated  that  currents  in  the  shallow  lagoon would be wind-driven and mainly
westerly.  Tides are about 10cm in range whereas storm  surges  are  often  greater
than  1m.   River runoff occurs during the very short summer season with 85% of the
annual flow occurring over a 10-day period in early June.  Rivers overflowing  the
shorefast ice can be seen along the whole coastline on satellite photographs.

## Physical contribution

The research associated with the physical aspects of the study was  undertaken
by  the author and his colleague Dr. J. C. H. Mungall.  During the initial workshop
we reprogrammed an existing model which has been described elsewhere (Mungall  and
Matthews  1978)  for  the  Prudhoe Bay-Simpson Lagoon coastal margin.  Steady state
wind conditions were applied for 4 wind velocities, 2.5m/s, 5m/s, 7.5m/s and  10m/s
from  ENE  and  WNW.   Tides were assumed to be negligible.  Using records of winds
during the open water season for a previous year, a typical series of wind  regimes

224

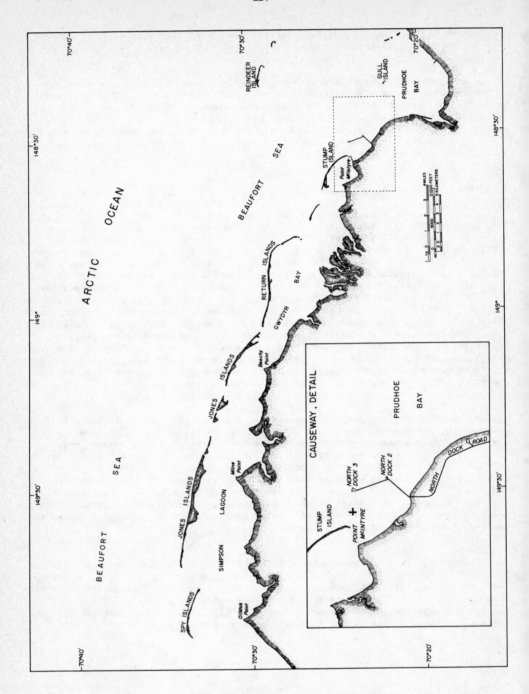

Figure 1. The study area and instrument location (+).

was used to construct a data series of mean daily flow rates through the lagoon system. These crude estimates were then used by the ecosystem simulation model. Preliminary estimates indicated that water current velocity would be 3% of the wind speed.

With the results of the simulation model we were able to determine which processes needed further field research. The physical oceanographic work involved verifying the predicted currents and sea level changes and determining the temperature and salinity in different parts of the lagoon system. In addition, as physical scientists, we were asked to provide hydrological and meteorological data. For these latter activities we enlisted the aid of colleagues. A preliminary field season was planned in which three current meters and two tide guages were deployed. Simultaneously 10m winds and surface barometric pressures were recorded at 3 island sites and 3 mainland sites. This was a period for the development of techniques. The second field season was designed to provide major data from all disciplines. The second season has been successfully completed at this writing but data analyses are not yet available. Results and discussion will be restricted to the first season.

Results

A tide gauge and current meter were moored between Stump Island and the ARCO North Dock 3 (fig. 1). This location was chosen because model results had indicated that this channel serves as a major source of water for Simpson Lagoon. The instruments were mounted in a specially-designed frame which rested on the bottom in 2.3m of water with the sensors 1.7m below the surface. Figure 2 shows hourly averages of data sampled at 5-minute intervals for salinity, temperature and current at this location. Also shown are wind data for Cottle Island, the long island off Beechey Point in figure 1. The Cottle Island weather station was installed specially for these studies because the government weather station at Prudhoe Bay airport is too far inland to be representative of the coastal weather conditions.

The most notable feature of the salinity plot (fig. 2) is the sharp drop in salinity between the August 16 and 18. The salinity fell from 31.86°/oo to 26.13°/oo in the 5-minute sampling interval between 0940 hours and 0945 hours local time on 16 August. At the same time the temperature sensor showed a steady rise from 1.11°C to a peak of 6.87°C at 0115 hours on 17 August. The subsequent salinity increase was equally sharp with the change from 25.46°/oo to 31.11°/oo in the interval from 0155 hours to 0200 hours local time on 18th August.

Current vectors indicate that currents towards the south and east predominate and correspond to lagoon inflow as anticipated. The current directions appear to follow the wind directions very closely with a lag of up to 3 hours. Current magnitudes are about 3% of the wind speeds as anticipated.

Figure 3 is the result of a helicopter survey of surface temperatures and salinity values flown by Dr. Mungall on August 15 1977 between the hours 1030 and 1925 local time. Note that warmer temperatures and lower salinities are generally found near the mainland coast. The range of salinity (14.7°/oo to 31.0°/oo) from the east and west sides of the ARCO dock should also be noted.

Figure 4 is an enhanced infra-red photograph of the arctic coast of Alaska taken 14 August 1977. The surface temperature has been split into two temperature

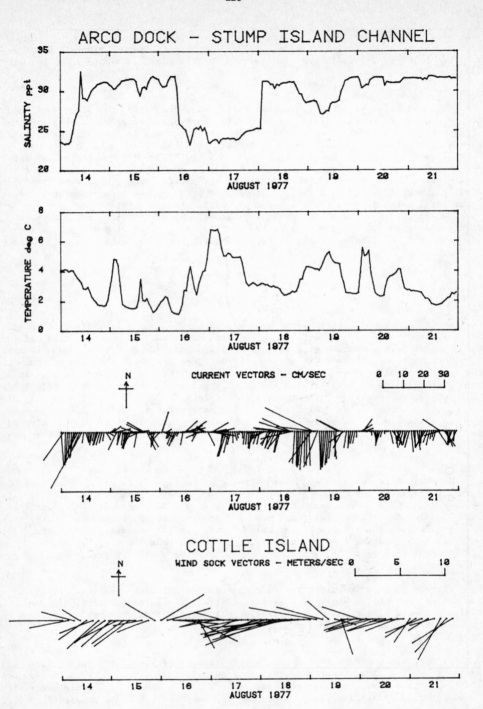

Figure 2. Hourly averages of 5-minute samples of temperature, salinity and current speed and direction for the ARCO dock - Stump island channel, and three-hourly wind sock vectors for Cottle Island.

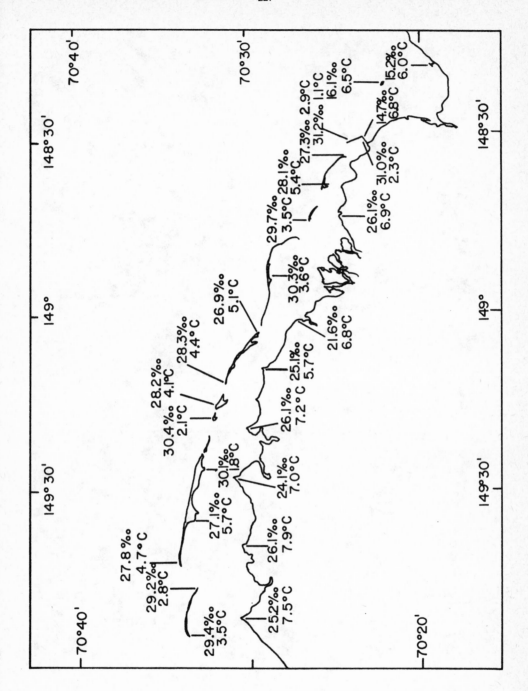

Figure 3. Salinity and temperature values 30cm. below the surface, at the coastlines, between 1000 hours and 1930 hours local time on 15 August 1977.

Figure 4. Infra red photograph of the Beaufort Sea coastline taken at 0645 hours local time on 15 August 1977.

ranges, -2.2°C to 6.5°C and 7.0°C to 15.5°C, which are shown as white-black ranges in the photograph. Nearly everywhere land is warmer than 15.5°C so that it appears black. Clouds colder than -2.2°C appear white in the photograph. The sea surface temperature ranges from near 0°C (grey) at the ice edge to 6.5°C (black) near the coast. A band of surface water warmer than 7°C (white) is clearly seen lying along the coast. Detailed examination of the original negative of the photograph shows a mass of water warmer than 7°C in Prudhoe Bay. The reslolution of the saltellite sensors is about 1km on the surface of the earth.

Currents at the Stump Island channel (fig 2) average about 8cm/s into the lagoon from the time of the satellite overflight to the appearance of the salinity drop observed on 16 August. The warm water mass observed in the photograph appears to be about 5km from our sensors. Water travelling at 8cm/s over a distance of 5 km would take about 18 hours in transit. The satellite overflight occurred some 15 hours before the salinity excursion was observed. It seems conclusive that the water mass observed by the satellite in Prudhoe Bay is the same one appearing on the temperature and salinity records as a water mass of low salinity and high temperature. Salinity is probably a better indicator of surface conditions in this case since the warm, brackish water is undoubtedly at the surface. The subsurface temperature sensor on the instrument package is 1.7m below the skin temperature seen by the satellite. Other areas of warmer temperature can be seen in the satellite photograph, especially off river mouths but also at other locations close to the coast. The data in figure 4 confirm the existence of a warm brackish water mass in Prudnoe Bay and suggest that other water masses are present along the mainland coast in Simpson Lagoon.

## Discussion

The results outlined above were used as verification data for the wind-driven numerical model. They gave confirmation of model predictions that currents were about 3% of the wind speed and in the direction of the wind. Currents did not appear to be much diminished through the channel between Stump Island and the ARCO causeway in numerical model runs with the causeway, compared with runs without the causeway.

We had not anticipated the large fluctuations in temperature and salinity observed at our instrument site. These data suggested that brackish, warm water was moving eastwards along the shoreline between patches of cold saline waters. The salellite infra-red pictures confirmed that water with surface temperatures warmer than 7°C could be found all along the arctic coast. Larger areas occurred off river mouths; a large patch observed in Prudhoe Bay appeared to originate in the Sagavanirktok River to the east of the bay. The sudden salinity drop and temperature rise observed between August 16 and 18 at our instrument site, is undoubtedly the water mass observed on the satellite photograph taken several hours earlier. Data taken by Dr. Mungall confirmed that the warmer, brackish water reached the east side of the causeway before our instruments detected it to the west. Measurements of temperature on the west side of the causeway were always more saline than on the east side suggesting that the causeway was tending to entrain saline ocean water in the leeward eddy.

## Ecological synthesis

The foregoing results and discussion were presented at an ecological meeting with other scientists. Fisheries biologists had observed that fish catches from gill nets were larger along the mainland shore than along any other shores or in lagoon or open waters. Moreover the catches nearshore showed great variation from

day-to-day. Fish had been observed swimming in water so shallow that their dorsal
fins protruded above the water surface. Unfortunately fish catch data were
available for daily periods of 24 hours. The physical data and fishery data suggested that fish preferred warmer and less saline waters.

A full hydrodynamic model employing density calculations might have predicted
the observed frontal systems. However such models are still in the developmental
stages and very expensive to run. This is clearly an area for further research effort, however, since such frontal systems are a common feature of estuarine
regimes. It was fortunate that that we had used a wind-driven numerical model to
discover verification stations - "points of influence", a term coined by Professor
Walter Hansen. As a result of this we did have instruments capable of measuring
temperature and salinity fluctuations in addition to the current sensors.

It was suggested from these results that impacts which broke up the flow of
warm water near the shore may interrupt the fish migrations which occur annually.
Fish are believed to winter over in fresh water pools in rivers and move to lagoons
to feed and breed each spring. The ARCO causeway which had previously been
modelled to examine its effect on current and sedimentation patterns might now be
responsible for more important salinity and temperature changes.

Although a numerical model of salinity and temperature fluctuations would be
useful, it was incompatible with the paucity of verification data, especially time-
series biological data. However we redirected the physical oceanography program to
take more detailed temperature and salinity measurements both temporally and spatially, in lieu of a detailed hydrodynamic modelling experiment.

The current measurements confirmed the model prediction that the lagoon was
flushed over a period of a few weeks and that it was, therefore, not an isolated
biological habitat. Biological sampling, which had concentrated on making cross-
section measurements in one place in the lagoon, was modified to look at two or
more cross-sections. This would allow the detection of fish migrating in a patch
of warm, brackish water.

Conclusions

The contribution of hydrodynamic modelling techniques to ecosystem process
modelling and to the decision-making process is proven. Our contribution showed
that the lagoon was not a lake-like habitat, but had a flushing period measured in
tens of days and thus was more like a riverine environment. This led to changes in
the biological sampling program.

Our prediction that the ARCO causeway did not greatly influence the circulation in the lagoon was confirmed. However the field measurements suggested that
the causeway could influence the temperature and salinity regime downstream of the
obstruction. Using the ecosystem process approach this new information was used
immediately to identify its impact on the local commercial fish species.

These results were used to modify the second field season program to take more
detailed measurements of these processes. Additionally the State of Alaska has
been alerted to the possible environmental effects of the ARCO causeway and is pur-
suing this problem. Restrictions have been put into effect which will stop all
oil-related activity between April and September so that the lagoon lease areas
would not be at risk during their highly productive open water season. Finally the

decision-makers, apparently impressed by the productivity of Simpson Lagoon as demonstrated by the ecosystem study, withdrew it from the proposed lease sale area.

We conclude that we have made and are making positive contributions by using hydrodynamic models as part of an ecosystem process study. The method demands flexibility of the scientist, but produces results rapidly and in a form useful to decision-makers. It is believed that this is a method worthy of further development.

References

Mungall, J. C. H. and J. B. Matthews, 1978
    The $M_2$ Tide of the Irish Sea: Hourly Configurations of the Sea Surface and of the Depth-mean Currents
    Estuarine and Coastal Marine Science, 6 : 55-74

# SALINITY INTRUSION MODELS

K. Fischer

Chair of Fluid Mechanics
Technical University Hannover
Federal Republic of Germany

## 1. Separation of Scales

In many cases, the salinity intrusion processes of estuaries are domin-
antly governed by the amount of fresh water discharge, the density
difference between fresh water and sea water, and the tidal amplitude.
The time for reaching periodic tidal conditions is of the order of one
day, while the time for reaching equilibrium conditions for the salinity
distribution is of the order of several weeks. Due to this fact, a time
averaging over the tidal motions can be performed in order to study the
salinity intrusion processes separately. The stream velocities of the
tidal motion are often of the order of 1 m/s, while the stream velocities
of the fresh water discharge is of the order of 1 cm/s. The deviations
in the vertical velocity profiles from the usual logarithmic shape which
are essential for the salinity intrusion process, are caused by the
horizontal density gradients; these deviations are of the order of
1 cm/s too [1]. So, if one is only interested in the tidal dynamics
of an estuary, the barotropic assumption of constant density is well
justified. Furthermore, if the salinity distribution in an estuary at
a certain time is known, the variations in salinity due to the tidal
motions can be calculated to a good degree of accuracy for a few tidal
cycles under the barotropic assumption, because the tidal and salinity
time scales are so different.
The most important aspect for the salinity intrusion problem, however,
is the calculation of an equilibrium state, and its dependency upon the
fresh water discharge; this can be done most economically by using tidal
averages. The remaining problem is then how to quantify the tidally-
induced mixing; this can be regarded as a closure problem, similar to
the closure problems in turbulence theory. In the following, all quantities and considerations apply to the tidally averaged case, if not
explicitly stated otherwise.

## 2. Simple Mathematical Models

The mathematical models which make use of the partial differential equations describing the physics of the salinity intrusion process can be classified according to their simplifying assumptions. There are models with spatially averaged variables, where the averaging is done in the vertical, or lateral, or both directions. Then there exist the two-(or more)-layer-models with constant or variable density in the different layers, and finally the models that are three-dimensional in space. Because of their relative simplicity, two types of models are used most widely, and shall be discussed in the following, i.e. one-dimensional models, and two-layer-models.

### 2.1 One-dimensional Model (1-d-Model)

The differential equations for models which are averaged in the vertical and lateral directions are [3]:

the equation of continuity of water

$$\frac{\partial h}{\partial t} + \frac{\partial}{\partial x} (h\,u) = 0 \tag{1}$$

the equation of motion

$$\frac{\partial}{\partial t} (h\,u) + \frac{\partial}{\partial x} (h\,u^2) + g\,h\frac{\partial h}{\partial x} + \frac{g h^2}{2\rho}\frac{\partial \rho}{\partial x} + \frac{\tau_b}{\rho} = 0 \tag{2}$$

and the equation of continuity of salt

$$\frac{\partial}{\partial t} (h\,\rho) + \frac{\partial}{\partial x} (h\,\rho\,u) = \frac{\partial}{\partial x} (h\,A\frac{\partial \rho}{\partial x}) \tag{3}$$

Here, h is the total water depth, u the velocity, $\rho$ the density, x the space-coordinate along the estuary axis, t the time-coordinate, $\tau_b$ the bottom friction, g the gravitational acceleration, and A the apparent diffusion coefficient. The estuary width was taken as constant, and the bottom slope zero.

Such a model is usually applied when the vertical stratification is low [4]. The main contributions to the parameter A are due to dispersive fluxes, caused by deviations of the actual vertical and lateral velocity profiles from the average value. Other contributions, like turbulent or molecular diffusion, can usually be neglected in this approximation [5]. It should be noted, however, that the numerical solution of (3) often leads to diffusion-like numerical errors which are comparable to the r-h-s of (3) [6]. This fact can make it difficult to estimate the parameter A from a comparison of measured and calculated salinity data.

## 2.2 Two-Layer-Model (2-1-Model)

The differential equations for the upper (index 1) and lower (index 2) layer are [7]:

the equations of continuity

$$\frac{\partial a_1}{\partial t} + \frac{\partial}{\partial x} (a_1 u_1) = 0 \tag{4}$$

$$\frac{\partial a_2}{\partial t} + \frac{\partial}{\partial x} (a_2 u_2) = 0 \tag{5}$$

and the equations of motion

$$\frac{\partial}{\partial t} (a_1 u_1) + \frac{\partial}{\partial x} (a_1 u_1{}^2) + g\, a_1\, \frac{\partial h}{\partial x} + \frac{\tau_i}{\rho_1} = 0 \tag{6}$$

$$\frac{\partial}{\partial t} (a_2 u_2) + \frac{\partial}{\partial x} (a_2 u_2{}^2) + (1 - \varepsilon) g\, a_2\, \frac{\partial h}{\partial x} + \varepsilon\, g\, a_2\, \frac{\partial a_2}{\partial x}$$
$$+ \frac{\tau_b - \tau_i}{\rho_2} = 0 \tag{7}$$

Here,

$$\varepsilon = \frac{\rho_2 - \rho_1}{\rho_2}$$

is the density anomaly, a is the layer thickness, and $\tau_i$ the interfacial friction. Here, the assumptions of constant width and zero bottom slope are made again.

This type of model is usually applied in cases where the vertical stratification is high [8]. For cases with partial stratification, the model is often extended to work with variable densities in both layers; then two additional equations similar to (3) are used [9]. The latter model is more complicated and shall not be considered here. The main problem in the application of the simple 2-1-model is the specification of the interfacial friction $\tau_i$.

## 2.3 Comparison of the Models

By vertically averaging the 2-1-model equations, it is possible to make a direct comparison with the 1-d-model. This comparison shall be done in order to study the effects of stratification, and to discuss the problem of model choice. The vertical averaged state variables are:

the density

$$\rho = \frac{1}{h} \; (\rho_1 a_1 + \rho_2 a_2) \tag{8}$$

and the velocity

$$u = \frac{1}{h} \; (u_1 a_1 + u_2 a_2) \tag{9}$$

recalling that

$$h = a_1 + a_2 \tag{1o}$$

By summing up (6) and (7) one obtains the equation of motion

$$\frac{\partial}{\partial t} \; (h\,u) + \frac{\partial}{\partial x} \; (h\,u^2) + \frac{\partial}{\partial x} \; (\, h\,[u_2 - u_1]^2 \; \frac{[\rho_2 - \rho][\rho - \rho_1]}{[\rho_2 - \rho_1]^2} \; )$$

$$+ g\,h\frac{\partial h}{\partial x} \; (\; 1 - \frac{[\rho_2 - \rho]\,[\rho - \rho_1]}{\rho_2 [\rho_2 - \rho_1]} \; ) + \frac{g\,h^2}{\rho} \; \frac{\partial \rho}{\partial x} \; \frac{\rho - \rho_1}{\rho_2 - \rho_1}$$

$$+ \tau_i \; \frac{\rho_2 - \rho_1}{\rho_1 \, \rho_2} + \frac{\tau_b}{\rho_2} = 0 \tag{11}$$

This expression has to be compared with (2). The third term in (11) is a correction to the convective acceleration term, proportional to the "shear velocity", $u_1 - u_2$, squared. It should be mentioned that the convective term in the 1-d-model equation (2) is only an approximate expression for the true vertical average of the three-dimensional convective term; essentially corrections like the third term used in (11) have been omitted, because they cannot be represented by the vertically averaged variables. As the "shear velocity" is of the same order of magnitude as the average velocity (and this is true for stratified and mixed estuaries as well), the representation of the convective term in the 1-d-model is poor. By the same reasoning it can be shown that the representation of this term in the lower layer of the 2-1-model is also poor, because in this layer usually a strong circulating flow with near zero mean values occurs. In many cases, however, the convective term is negligible against the other terms in the equation of motion.

The fifth term is the baroclinic pressure term; it is equal to the 1-d-model term, when the interface is in the middle of the water column (or the density $\rho$ in the middle between $\rho_1$ and $\rho_2$). If the interface is deeper or the density lower, the 2-1-model has a smaller baroclinic pressure term than the 1-d-model, and vice versa. As this term is the driving force for the salinity intrusion, it can be seen that vertical mixing near the seaward end of the estuary can reduce the salt intrusion, as compared to stratified conditons. The remaining corrections (part of fourth term, sixth term) are negligible.

A convection - diffusion equation similar to (3) can be obtained by forming the weighted sum of (4) and (5), where the weights are the respective densities of the layers:

$$\frac{\partial}{\partial t}\,(h\,\rho) + \frac{\partial}{\partial x}\,(h\,\rho\,u) = \frac{\partial}{\partial x}(h\,[u_1 - u_2]\,\frac{[\rho_2 - \rho][\rho - \rho_1]}{\rho_2 - \rho_1}\,) \tag{12}$$

By comparison with (3) the following expression for the dispersion coefficient is obtained:

$$A = \frac{1}{\frac{\partial \rho}{\partial x}}\,(u_1 - u_2)\,\frac{(\rho_2 - \rho)\,(\rho - \rho_1)}{\rho_2 - \rho_1} \tag{13}$$

This expression is directly comparable to formulae cited by other authors [1o], where the "shear velocity" is replaced by the fresh water velocity $u_1$.

Now it is readily obvious that the uncertainty in choosing parameters is the same for the 1-d-model as for the 2-1-model. For the 1-d-model, choosing a value for the dispersion coefficient A is equivalent to using equ. (13), and choosing a value of the "shear velocity" $u_1 - u_2$, because the density values $\rho(x)$, $\rho_1$ (fresh water) and $\rho_2$ (sea water) are known. For the 2-1-model, choosing a value for the interfacial friction is equivalent to choosing a value of the "shear valocity", because in most cases expressions like

$$\tau_i = \lambda \rho_1 |u_1 - u_2| (u_1 - u_2) \tag{14}$$

are used [7], where $\lambda$ is a constant to be determined. To each value chosen for $\lambda$ there corresponds a certain solution for the "shear velocity" as expressed in (14), since the interfacial friction can be calculated from (6). Neglecting the first two terms in (6), and substituting the interfacial friction in (14), one obtains

$$\lambda = - \frac{g\,h}{|u_1 - u_2|(u_1 - u_2)} \frac{\partial h}{\partial x} \frac{\rho_2 - \rho(x)}{\rho_2 - \rho_1} \qquad (15)$$

where the analogy to (13) is evident.

There exists a unique relation between the 1-d-model parameters and the 2-1-model parameters. With the given density values $\rho_1$, $\rho_2$, and $\rho(x)$, and the depth $h(x)$, the layer thicknesses are

$$a_1 = h \frac{\rho_2 - \rho}{\rho_2 - \rho_1} \quad \text{and } a_2 = h \frac{\rho - \rho_1}{\rho_2 - \rho_1} \qquad (16)$$

Furthermore, equations (9) and (13) or (15) can be used to calculate the layer velocities $u_1$ and $u_2$.

By elimination of the "shear velocity" in (13) and (15), an explicit relation between the coefficients A and $\lambda$ is obtained as follows:

$$A = ( \frac{\rho_2 - \rho}{\rho_2 - \rho_1} )^{3/2} \frac{\rho - \rho_1}{\frac{\partial \rho}{\partial x}} \sqrt{|\frac{g\,h}{\lambda} \frac{\partial h}{\partial x}|} \qquad (17)$$

It should be noted that by such an approach the solution for free wave motion of the interface cannot be obtained. The reason for this is found in the fact that only three dynamical equations are solved in the 1-d-model, while there are four equations in the 2-1-model. A fourth equation which could be added to the 1-d-model is an equation of motion for the "shear velocity", but for the salinity intrusion problem it seems to be justified to replace such an equation by (14), which means that the internal motions are dominated by friction (and not by inertia).

As a consequence, it is unnecessary to use 2-1-models for stratified flow, because one can use the simpler 1-d-models equally well. The only modification in the 1-d-model that takes account of the stratification is the different baroclinic pressure force term; all other corrections in (11) are less important. One could use a linear combination of the "mixed" and the "stratified" baroclinic terms in order to have a unified model:

$$g\,h^2 \frac{\partial \rho}{\partial x} [\frac{\alpha}{2\rho} + \frac{1 - \alpha}{\rho_2} \frac{\rho - \rho_1}{\rho_2 - \rho_1}] \qquad (18)$$

where $\alpha = 1$ for the fully mixed, $\alpha = 0$ for fully stratified, and inter-
mediate values for partly mixed cases. The value of $\alpha$ can be calculated
from a measured vertical density profile as follows: First calculate
the vertical average density $\rho$, then the layer thicknesses by (16),
then the average in the upper layer

$$\bar{\rho}_1 = \frac{1}{a_1} \int_{a_2}^{a_2+a_1} \rho\,(z)\ d\,z \tag{19}$$

and finally the coefficient

$$\alpha = \frac{\bar{\rho}_1 - \rho_1}{\rho - \rho_1} \tag{2o}$$

## 2.4 Interfacial Friction

As was shown in the preceding section, the baroclinic pressure force
term is of the same order of magnitude for the stratified and mixed
model. Now an estimate will be given for the deviation of the interfacial friction from its barotropic value ($\varepsilon = 0$). In the latter case,
we obtain from (6), by neglecting the time derivative and the convective term

$$\tau_{i,o} = \tau_i(\varepsilon = 0) = -\rho g\, a_1 \frac{\partial h}{\partial x} \tag{21}$$

This is the well-known linear dependency of the vertical coordinate represented by $a_1$.

The difference of (6) and (7) is used to calculated $\tau_i$ for the baroclinic
case (time derivatives and convective terms are neglected). We obtain,
for the relative deviation of $\tau_i$ from its barotropic value

$$\frac{\tau_i - \tau_{i,o}}{\tau_{i,o}} = \frac{\rho - \rho_1}{2\rho} + \frac{h}{2\rho}\ \frac{\frac{\partial \rho}{\partial x}}{\frac{\partial h}{\partial x}}\ (\,1 - \frac{\rho - \rho_1}{\rho_2 - \rho}\,) \tag{22}$$

The first term on the r. h. s. of (22) is small against 1 and can be
neglected. The second term is zero for $\rho$ in the middle between $\rho_1$ and
$\rho_2$. From (2) follows that the second term is of the order of 1 when $\rho$
is different from the middle value. This means that the interfacial
friction and the vertical velocity profile are strongly changed by the
horizontal density gradient, irrespective of the degree of vertical
stratification. The "shear velocity" is of the same order of magnitude

as the fresh water velocity. An increased mixing does not remove the necessity to consider the force balance between baroclinic and barotropic terms in order to obtain an adequate salinity intrusion model.

The choice of the interfacial friction coefficient, or, equivalently, of the longitudinal dispersion coefficient, is of central importance. It is known that these coefficients depend, in a complicated way, on the Reynolds and Froude numbers and the stratification of the estuary [11], but for practical purposes one usually has to adjust these coefficients by comparison of calculated and measured salinity data [8]. This means that the models considered until now are too schematic for purposes of prediction and extrapolation, and should be replaced by more elaborate ones. A special example for the over-schematization of the 2-1-model is given in the following:

Consider a closed basin filled with two layers of water of different density [12]. The steady-state circulation when wind stress is applied to the surface is shown schematically in figure 1.

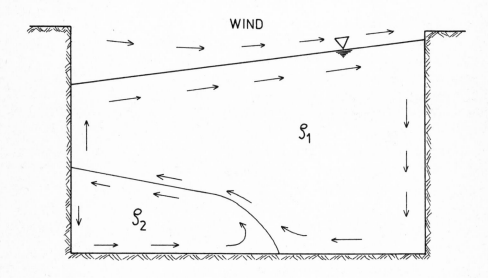

Fig. 1: Wind stress on a closed basin, filled with a two-layer fluid

The surface slope is balanced by the wind stress. The interfacial slope should be balanced mainly by the interfacial stress. However, in the 2-1-model, under steady-state conditions, the velocities $u_1$ and $u_2$ are zero, so the interfacial friction vanishes, and the model is unable to reproduce the wedge shape of the interface like in figure 1. The same bad result is obtained when a multi-layer model is applied.

## 3. Models of Dispersive Fluxes

The main shortcoming of the simple mathematical models described above is their necessity to parametrize the dispersive fluxes through the choice of the interfacial friction parameter, or the effective dispersion coefficient. As these fluxes are eddies of a size comparable to the spatial model extensions, they can be modelled in a consistent fashion with the other model variables. A vertical model discretization which allows for the representation of the actual velocity profiles (including recirculations like in the wedge of figure 1) can be used to simulate the dispersive flux driven by the longitudinal density gradient. Estuaries with shallow lateral areas show strong transversal circulating dispersive fluxes; they can be modelled by means of a lateral discretization. Instead of the dispersion parameter in the 1-d-model, turbulent friction and diffusion parameters are to be used now, and it can be shown that the vertical eddy viscosity coefficient is the most important one for salinity intrusion problems. The step from the 1-d-model to the vertically and laterally discretized model corresponds to the rising of the turbulence closure by one level. The coefficients applied on this level are much more specific and therefore less variable than the coefficients of the lower level, and mixing-length approximations may be sufficient in many cases. Of course, explicit turbulence models can be used to reach the next level of closure, if this is necessary.

It should be noted that the turbulent diffusion parameters are, in general, two orders of magnitude smaller than their dispersive counterparts, and often they are negligible (e.g. the horizontal diffusion terms are sometimes smaller than similar numerical errors).

As a result, salinity intrusion models for estuaries should be three-dimensional in space. This permits a consistent representation of all relevant mechanisms in the same scale of space and time, as average flow velocity, water level, density, and internal fluxes, while effects of different scales should be parametrized or neglected. In spite of the practical problems with such models, this should be a reason for continuing and intensifying the efforts in development and application of such models.

## 4. References

1. A. T. Ippen, "Salinity Intrusion in Estuaries", in [2], chapter 13.

2. A. T. Ippen (ed.), "Estuary and Coastline Hydrodynamics",
   McGraw-Hill, New York (1966)

3. G. Abraham, "Density Currents Due to Differences in Salinity",
   Rijkswaterstaat Communications 26, Delft (1976)

4. M. L. Thatcher, D. R. F. Harleman, "A Mathematical Model for the
   Prediction of Unsteady Salinity Intrusion in Estuaries",
   R. M. Parsons Laboratory, Technical Report No. 144,
   M. I. T.,(1972)

5. K. Sanmuganathan, C. L. Abernethy, "A Mathematical Model to Predict
   Long Term Salinity Intrusion in Estuaries", Proceedings,
   2nd World Congress, International Water Resources Association,
   p. 313, New Delhi (1975)

6. J. E. Fromm, "A Method for Reducing Dispersion in Convective
   Difference Schemes", Journal of Computational Physics 3,
   176 (1968)

7. C. B. Vreugdenhil, "Computation of Gravity Currents in Estuaries",
   Delft Hydraulics Laboratory, Publication No. 86 (197o)

8. R. Dazzi, M. Tomasino, "Mathematical Model of Salinity Intrusion
   in the Delta of the Po River", Proceedings 14th Coastal
   Engineering Conference 3, p. 23o2, Copenhagen (1974)

9. J. P. Grubert, M. B. Abbott, "Numerical Computation of Stratified
   Nearly Horizontal Flows", Journal of the Hydraulics
   Division, ASCE, 98, p. 1847 (1972)

1o. D. R. F. Harleman, "Diffusion Processes in Stratified Flow", in
    [2], chapter 12

11. K. Lofquist, "Flow and Stress Near an Interface between Stratified
    Liquids", The Physics of Fluids 3, p. 158 (196o)

12, B. Hellström, "Wind Effects on Lakes and Rivers", Ing. Vet. Akad.
    Handl. 158 (1941)

# A POINT OF VIEW: PHYSICAL PROCESSES ON THE
## CONTINENTAL SHELF AND THEIR IMPLICATIONS
### FOR NUMERICAL CIRCULATION MODELS

Christopher N. K. Mooers
College of Marine Studies
University of Delaware
P. O. B. 286

Lewes, Delaware  19958/U.S.A.

When contemplating numerical circulation models for continental shelf regimes, it
is necessary to have a clear view of the prevailing processes, and their properties.
Such regimes are challenging because they are under the joint influence of density
stratification, variable depth, earth's rotation, and bottom friction.  Further
factors are the strong response of shelf regimes to winds and the general vigor of
tides and river run-off in these regimes.  The approach taken here is to make a conceptual Fourier decomposition so that the processes can be ordered in Fourier space.

At frequencies greater than a few cycles per day, there are, of course, turbulence,
surface gravity waves, and internal gravity waves.  In circulation models, the effects
of these processes would generally be parameterized.  Though both surface and internal gravity waves generally have offshore sources, they can also exist as coastally-
trapped wave modes, causing radiation condition problems in a model of finite alongshore extent.

In the 1 to 2 cycles per day band at mid-latitudes, near-inertial as well as tidal
motions are quite prevalent on continental shelves, especially when the water column
is density stratified.  Then near-inertial motions generated by favorable changes
in the wind can propagate downwards through the water column, as well as deepen the
surface mixed layer through shear instabilities and subsequent turbulent entrainment.
Under stratified conditions, internal tides generated near the shelfbreak propagate
shoreward.  They, too, can produce shear instabilities while dissipating on the
shelf.  Tidal currents and bottom friction also produce turbulence necessary for so-
called tidal-stirring of the lower (sometimes, all of the) water column.  In this
band and over continental shelves, motions generated by the diurnal sea breeze may
need to be taken in account due to their influence on mixing.  In a numerical circulation model, it may be very desirable to simulate the tidal, near-inertial and
diurnal motions, at least to the extent that they govern mixing.

In the 0.1 to 1.0 cycles per day band, the dominant motions are those generated by
synoptic scale atmospheric disturbances (cyclones, fronts, etc.).  In seeming
contrast to the open ocean, the response of shelf waters to these disturbances is
very intense.  In addition to the familiar setup (storm surge) or set down, these
disturbances generate transient coastal upwelling and coastally-trapped topographic
Rossby waves.  These waves are essentially vorticity waves; as such, they are

strongly manifested in horizontal currents.  Since they propagate alongshore (cum sole only), they also cause radiation condition problems for models of finite alongshore extent.  Except for the tidally-dominated shelves, motions on these time scales generally produce the most vigorous circulation on continental shelves; hence, they are of significance to sediment transport, pollution, dispersal, and primary productivity processes.

In the 0.1 to 1.0 cycles per month band, the dominant motions on the shelf are those generated by oceanic eddies, and perhaps planetary Rossby waves, impinging upon the continental shelf.  In this regard, there may be a substantial asymmetry between shelf circulation on the eastern and western sides of ocean basins because long planetary Rossby waves propagate energy westward.

In the 0.1 to 1.0 cycles per year band, the annual cycle of stratification and circulation, and their interannual variability, prevail.  In this band, if nowhere else, the effects of thermohaline forcing must be taken into account.  It is here too where the quality of our predictive knowledge may be most sensitive for fisheries and water quality interests.

This account in Fourier space considered only the frequency coordinate and not the wave number coordinates.  Due to the bounded and singular nature of shelf regimes, it may be preferrable to consider localized spatial domains, or to possibly use cross-shelf normal modes and a Fourier decomposition in the alongshore coordinate. For example, special treatment is needed for the dynamics of surface and bottom mixed layers.  Also, particular care is needed with circulation models in the proximity of various mesoscale topographic domains:  submarine banks, canyons, basins, and capes, as well as coastal capes and embayments.  The nearshore boundary is singular with its breaking waves, alongshore jets, and riverine and estuarine run-off plumes and fronts.  The shelfbreak regime is complex because oceanic fronts; exchanges between shelf and oceanic waters; and baroclinic currents, undercurrents, and countercurrents occur there.  Also, an oceanic boundary current is generally just offshore, and meanders of, and spin-off eddies from, such currents often impinge on the shelfbreak.  Major coastal capes are separated by ca. 1,000 km and serve to isolate, to some degree, the circulation in contiguous shelf regimes. Hence, it is tempting to consider modeling such regional regimes in isolation, yet, there are persistent flows or at least episodes of intense transfers around such capes which must be treated.  In all of these problems, the barotropic and baroclinic radii of deformation enter as fundamental length scales.  Since the former is of the order of a 1,000 km (much larger than the shelfwidth) while the latter is of the order of 30 km on a shelf, then important differences in baroclinic circulation can be expected due to whether a shelf, submarine bank, or submarine canyon is wider or narrower than a baroclinic radius of deformation.  The corresponding oceanic scale determines the radius of spin-off eddies and widths of boundary currents, etc. (Similarly, the time scales of processes can be usefully sorted by whether they are smaller or greater than an inertial period.)

When the rich spectrum of processes outlined above has been considered, it is clear that operational shelf circulation models will require accurate bathymetry (perhaps with small scale features, too); an excellent climatology of currents, hydrography, run-off, winds, waves, solar heating, etc.; process "inventories", models, and parameterization experiments; and a continuing (probably real-time) data network to determine initial and boundary conditions and to make verifications. The data requirements may not be horrenduous if fairly careful sampling and process experiments are conducted before establishing monitoring networks. On the other hand, sensitivity testing of numerical shelf circulation models is needed to set priorities for process studies. Finally, it is abundantly clear that climatologies, process studies, and numerical circulation models must now be developed together on continental shelves; it would be interesting to see such coordinated development done in at least one shelf regime.

The following is an abstract of Dr. Magnell's presentation at the Symposium on "Mathematical Modelling of Estuarine Physics". The work is fully described in:

Bennett, J. and B. Magnell, 1979. A Dynamical Analysis of
    Currents near the New Jersey Coast, Journal of
    Geophysical Research, 84, (C3), February 1979


OBSERVATIONS OF CONTINENTAL SHELF CIRCULATIONS
AND THEIR RELATION TO
MODEL VERIFICATION AND APPLICATION

Dr. Bruce Magnell
EG&G, Environmental Consultants
Waltham, Massachussetts


Current meter and wind data have been obtained from a location about 4.5 km off the coast of New Jersey (USA) in about 13 meters water depth. The coast is straight and the topography is fairly simple in the area of the measurements. As expected, the observed non-tidal currents were nearly shore-parallel, and were generally dominated by wind stress and bottom friction. However, there were indications of significant non-local effects. To investigate these, Dr. John Bennett of the Massachussetts Institute of Technology (now at the Great Lakes Environmental Research Laboratories) developed and applied a two-dimensional numerical model (vertical and offshore) of wind-driven coastal circulation. The model neglects longshore variations of current and all variations of density, but includes inertial accelerations and a non-linear eddy viscosity. Local wind stress, sea level changes, and a constant longshore pressure gradient are the forcing terms.

Our purpose was to model the local forcing as well as possible, and by comparing the model currents with the observations, to learn the nature of the non-local effects. By comparison with a fully three-dimensional model, this approach was much simpler and better suited to the observations; and the numerical model results were relatively easy to understand and evaluate.

The model successfully reproduces most of the current variance; however, the predicted currents do not exhibit the dominant 4-hour response time of the observed currents, and the model infrequently misses energetic current events. We conclude that the effect of distant boundaries is to cause transient shore-parallel pressure gradients, which are opposed to the wind stress (a "set-up" effect). These pressure gradient forces build up slowly, and eventually limit the response of the current to the wind. Under certain circumstances, they can also cause relaxation currents of considerable amplitude.

We make no claim to have described the physics of the set-up effect. Rather, the significance of this work lies in the use of a combination of observations and a simple numerical model to demonstrate the existence of an important oceanic phenomenon.

# MATHEMATICAL MODELS OF SEDIMENT TRANSPORT
## IN CANALISED ESTUARIES

Dr. James G. Rodger
Estuary Division,
Hydraulics Research Station
Wallingford, UK

## INTRODUCTION

Rather than attempt to give a report on the whole state of the art of mathematical modelling of sediment transport in estuaries, a brief account of some of the models being used and developed at HRS will be given.  These mathematical models have all been developed for the pur- pose of predicting changes in the pattern of siltation in navigation channels or tidal basins as a result of altering the equilibrium of an estuary either by regulation of the fluvial flow or by engineering works. Most of the models have proved to be accurate enough for practical engineering problems especially for the comparison of different schemes in the feasibility stages of a project.

Because tidal processes in an estuary interact in a non-linear fashion, it is not possible to try to calculate the movement of sediment independ- ently of other tidal processes such as saline intrusion and vertical mixing.  A good model must be capable of simulating all the primary and secondary processes that effect scour, transport and deposition of sedi- ment in an estuary.  A sediment transport model can be considered as being composed of two main interacting parts: a hydraulic part and a sediment part.  Both parts of the model must be chosen to suit the estuary being considered.

In the present generation of estuary models, the sediment load is divided into two parts; the sand fraction moving as a function of the local instantaneous tidal conditions and the mud fraction moving fully suspend- ed in the flow.

The prediction of sand transport in canalised estuaries is easier than many other situations because the sediment fractions are usually very well sorted and the flux of all but the finest sand fractions (64 μm- 128 μm) passing a section are strongly influenced by local flow condi- tions and their proportions on the bed surface.  HRS find that the best results are obtained by using generalised semi-empirical functions fitted to data from the particular estuary under investigation eg sand transport, settling velocity and rate of mud scour.

The method employed by HRS at present calculates the transport of individual sediment fractions including mud and logs the local structure of the bed. This method allows the composition of the bed surface to vary from muddy to sandy, rough or smooth and permits armouring of fine sands by overlying layers of coarser sediment.

HYDRAULIC MODELS

Three different hydraulic models are currently of interest: an area averaged one-dimensional model, a two layer area averaged model and a two-dimensional laterally averaged model currently under development.

The results of any hydraulic model are dependent on the empirical relationship they use to calculate the effect of bed friction, dispersion and turbulent exchange.

One-dimensional models have been used for some time to model well mixed estuaries and require a coefficient of longitudinal dispersion. At present, for practical calculations, the coefficient of dispersion is often adjusted in the proving stages of the model. If the model is to be used in a predictive manner to evaluate the effect of engineering works then this method of tuning is not very satisfactory. At present a study is under way to try to obtain a generalised functional relationship for the coefficient of longitudinal dispersion for use in one-dimensional models.

The two layer model requires semi-empirical relationships to describe the interaction and mixing between the two layers. A mixing length approach has been used and a semi-empirical mixing function was obtained from field data (HRS 1974). This two layer model has now been applied successfully to several estuaries.

One of the most recent models being developed at HRS is a two-dimensional, laterally averaged model for application to partially mixed estuaries. This model uses the laterally averaged equations of mass, momentum and volume and requires expressions for the vertical exchange of mass and momentum in stratified conditions due to the turbulent fluctuations in the flow. A mixing length theory has been used and assuming the mixing length is known as a function of depth in homogeneous conditions the problem was to determine the mixing length for momentum and solutes in stratified conditions as a function of the flow variables and geometry of the channel being modelled. In the past, researchers used relationships of the form

$$l_m = l_o f(R_i)$$

where $l_m$ is the mixing length in stratified conditions and $l_o$ that for homogeneous conditions and f is a function of the local Richardson number. This relationship was examined using data collected in a straight canalised estuary where f was assumed to take the form

$$f(R_i) = (1 + \beta R_i)^n \qquad\qquad \beta, n = \text{constants}$$

Kent and Pritchard (1959) used this function for n = -1 and Rosby and Montgomery (1935) used it for n = -1/2. It was found that, in general, the momentum mixing length is not a function of the local Richardson number (Odd and Rodger, 1978). This is in agreement with Ellison and Turner's observation that, since the scale of turbulence covers a considerable height range, it was by no means obvious that there should be a simple dependence (of $l_m/l_o$) on a strictly local parameter such as the gradient Richardson number.

After analysing over 40 mixing length and Richardson number profiles it was found that when the Richardson number had a maximum below 3/4 of the total depth, the mixing length was uniform over most of the depth and the momentum mixing length was insensitive to $R_i$ values greater than 1. That is, the peak in the Richardson number had a limiting effect on the mixing length throughout the depth. The best fit to the data was obtained by using Rosby and Montgomery's relationship for $\beta = 160$ with the peak Richardson number and allowing this value of the mixing length to apply throughout the depth except where it was greater than the homogeneous mixing length. In cases where the Richardson number increased monotonically towards the surface or where the peak was close to the surface, Rosby and Montgomery's relationship gave good agreement using the local value of $R_i$.

The mixing length for solutes was found to be a function of the local $R_i$ and the relationship

$$\frac{l_c}{l_m} = \frac{1 - \dfrac{l_c}{l_m}\dfrac{R_i}{R_f}}{\left(1 - \dfrac{l_c}{l_m}R_i\right)^2}$$

gave good agreement with the observed data for $R_f = 0.08$.

Both these mixing relationships - for momentum and solutes - are required if the two-dimensional hydraulic model is to work successfully.

## SEDIMENT TRANSPORT

Typically, the sediments modelled in estuaries can be split into a mud fraction and a sand fraction.

The mud, if present, can be described by the equation of conservation of mud. In the two-dimensional model, this equation requires an expression for the vertical mixing due to turbulence and the mixing length for mud is assumed to be equal to that for solutes. Any model describing the movement of mud requires an empirical expression for the settling velocity often as a function of the concentration. At present, it is found that for each mud being modelled, the settling velocity is most accurately determined in the field (HRS 1976). The mud equation also contains a source/sink term describing deposition or erosion from the bed. The representation of this term depends on the type of mud being modelled and at present is determined from flume tests.

The sand fraction could be divided into a bed load and a suspended load. However, in order to model the movement of fine sand in the Gt Ouse a few years ago, a total load function was developed to describe the movement of individual size gradings of fine sand. This function was based on Bagnold's hypothesis that the rate of work done keeping sediment in suspension over unit bed area is related to the rate of dissipation of energy per unit area. (HRS 1973).

Since the movement of the tide over a long period has sorted the mud and sand according to size and degree of consolidation along the length of an estuary, any model of sediment transport must record the composition of the bed according to position and depth into the bed. To do this the bed is divided into layers and the composition of each layer is recorded. In this way the model can correctly simulate the situation where, as erosion occurs, layers of fine sand underlying layers of coarser sand are protected by this coarse sand from being eroded.

## BED LEVEL CHANGES

Normally, the change in bed level in one tide due to erosion or deposition is of the order of a few millimeters per tidal cycle. In order to examine the long term bed level changes due to altering the equilibrium of the estuary, the model would have to be run for several hundreds of tides which would be expensive and probably impractical.

So a method of accelerating the bed level changes in relation to the tidal calculation was developed initially for application to the Gt Ouse.

Briefly, the method involves dividing all the tides in a given period into classes (typically by tidal range) and multiplying the effect of each class by the number of times the tide occurs. For example, on the Gt Ouse, the effects of a thirty month drought were calculated by running a series of six tide types six times. This allowed changes in bed level to influence the flow which in turn affected the rate of sediment transport. Transient effects due to the shortened tidal cycle could have an unrealistic effect on the results so the type of estuary on which this method is used must be chosen carefully.

## BED FRICTION

When dealing with a mobile bed, the bed friction has always been a source of problems. An empirical expression (HRS 1974) was developed as a result of the field study of the Gt Ouse relating the bed friction factor or roughness length to the mobility number which, for conditions prevailing on the Gt Ouse, was independent of the rate of growth or decay of ripples. This relationship has been used successfully on other estuaries and will be used in the two-dimensional model of the Brisbane estuary.

## CONCLUSIONS

Empirical relationships are an important feature of any numerical model of sediment transport and their use is often far from being satisfying. Research, hopefully, will improve the representation of some of the physical processes currently described by these empirical formulae and may yield methods for eliminating the use of some of them completely (for example, the mixing length approach could be replaced by a turbulence transport model). However, the models described above are used in practical engineering problems and are therefore subject to constraints on cost and computer size available and the benefit obtained from a more sophisticated approach may not as yet be justified.

## REFERENCES

"Numerical Model Studies of the Gr Ouse", "A Transport Function for Fine Sand in the Estuary". Report DE 9, Hydraulics Research Station, Wallingford, United Kingdom, 1973.

"Numerical Model Studies of the Gt Ouse", "A Mixing Length Function for Vertical Exchange in Turbulent Stratified Two-Layer Flow". Report DE 11, Hydraulics Research Station, Wallingford, United Kingdom, January 1974.

"Frictional Resistance of a Sandy Tidal Channel. A Field Study". Report INT 125, Hydraulics Research Station, Wallingford, United Kingdom, June 1974.

"Determination of the Settling Velocities of Cohesive Muds". Report IT 161, Hydraulics Research Station, Wallingford, United Kingdom, October 1976.

Kent, R.E. and Pritchard D.W. "A Test of Mixing Length Theories in a Coastal Plain Estuary". Journal of Marine Research, Vol 1, 1959, pp 62-72.

Odd, N.V.M. and Rodger, J.G. "Vertical Mixing in Stratified Tidal Flows". Journal of the Hydraulics Division, ASCE, Vol 104, No HY3, March 1978.

Rosby, C.G. and Montgomery, R.B. "The Layer of Functional Influence in Wind and Ocean Currents". Pap. Phys. Oceanogr., Vol 3, No 3, 1935.

Ellison, T.H. "Turbulend Transport of Heat and Momentum from an Infinite Rough Plane". Journal of Fluid Mechanics, Vol 2, 1957, pp 456-466.

Ellison, T.H. and Turner, J.S. "Mixing of Dense Fluid in a Turbulent Pipe Flow". Journal of Fluid Mechanics, Vol 8, 1960, pp 514-544.

NUMERICAL MODELLING OF SEDIMENT TRANSPORT IN COASTAL WATERS

By Jürgen Sündermann and Walter Puls

Summary

The sediment transport in the German coastal zone of the North Sea is
of great importance for coastal engineering and navigation. Therefore
it's investigation and it's forecasting simulation is important, also
with respect to economical reasons.
In the following, two numerical models for tidal induced sediment
transport will be developed: one for the mesoscale (migration and deformation of dunes) and the other for the macroscale (long time development in coastal zones). The numerical results are compared with
measurements in hydraulic models and in the field.

## 1. Introduction

The sediment transport in tidal influenced coastal waters, estuaries
and rivers is an important and economically grave phenomenon. Navigation channels and coast protection are directly affected by sedimentation and erosion processes. Every constructional intervention in
the coastal zone changes the very sensitive balance of sedimentation
processes and can lead to irreversible morphological changes. So the
coastal engineer is required to analyze the sedimentation processes
and possibly to predict them.

However, the interactions involved in sediment transport are extremely complex. The driving forces are tide, wind, and wave motions,
which produce various velocity profiles. Baroclinic effects arise if
salinity or temperature differences occur. Ordinarily the flow is
turbulent. The transport of sediment takes place partly near the bed,
partly in suspension. Thus, we have a two phase flow, which has physical properties different from that of pure water. Sediment transport
changes the topography of the bottom, which in turn affects the flow.
Finally, a special difficulty is that hydrodynamic and sedimentologic
processes take place in different scales, which requires especially
sophisticated models with regard to computer economy. Up to now it is
not possible to involve all these complex interdependencies in an
integrated model in a quantitative way.

Therefore steps are necessary towards the numerical modelling of these
processes. Coastal engineers take great risks by coming to decisions
on coastal structures, which have far-reaching consequences for the
budget of sediment. Today there are available only some empirical
methods, which only give very rough data. It is the philosophy of the
model, developed below, that every mathematical model, however simple
it may be, gives additional objective help for a decision. It will
not be possible to describe the entire physics of sediment transport
completely. It is rather a question of a combined theoretical-empiri-
cal system of formulas, which is continuously verified by means of
field data, and which gives approximate quantitative, but detailed
information about rates of transport and resulting topographical
changes.

The models must be adapted to the scale in question. They will start
with simple formulations and will stepwise become more complicated.


## 2. Basic assumptions

It is assumed that sediment transport depends on the three dimensional
motion of water according to the shear stress concept of Shields. The
general interdependence between flow, transport and bed deformation is
shown in Fig. 1.

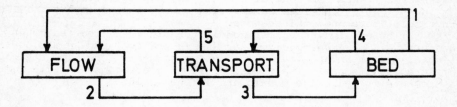

Fig. 1. The principal interactions in sediment transport


It is characterized by certain interactions that are marked with the
numbers 1 to 5 in the scheme. They refer to different scales and are
of different importance according to the scale in question. A hierarchy of models can be developed, where stepwise more and more interdependencies are involved.

The observation of sedimentation processes in the coastal zone leads
to a classification of these processes into spatial scales, which are
defined as follows:

$$L < 1 \text{ cm} \quad \text{microscale}$$
$$0.1 \text{ m} < L < 100 \text{ m} \quad \text{mesoscale}$$
$$L > 1 \text{ km} \quad \text{macroscale}$$

The macroscale characterizes global events in large coastal- or river
sections, the mesoscale includes the deformation and migration of
large bedforms, the microscale is not considered here.

According to the scheme of Fig. 1, the basis for further considerations
is the three dimensional field of water motion, which is determined
by the following differential equations:

Reynolds equation of motion

$$\frac{\partial v_i}{\partial t} + v_j\frac{\partial v_i}{\partial x_j} + \varepsilon_{ij} v_j = \frac{1}{\rho} (k_i - \frac{\partial p}{\partial x_i}) + \frac{\partial}{\partial x_j} (A_j\frac{\partial v_i}{\partial x_j})$$

Equation of continuity (1)

$$\frac{\partial v_j}{\partial x_j} = 0 \qquad (i,j = 1,2,3)$$

The notations are as follows:

$v_i$    component of the velocity vector in the $x_i$-direction
$p$     pressure
$k_i$    component of the external force in the $x_i$-direction
$\rho$     density of water
$\varepsilon_{ij}$   Coriolis tensor
$A_i$    eddy viscosity coefficient in the $x_i$-direction

In the cartesian co-ordinate system $\{x_i\}$, $x_1$ and $x_2$ are horizontal directions, $x_3$ is directed up vertically. The origin is in the undisturbed water surface.

The system (1) is completed by initial- and boundary conditions. Due to the nonlinearity of the equations and the complex geometry of natural areas, (1) must be solved numerically, for instance by a finite difference method.

As a solution one gets the spatial and temporal distribution of $v_i$ and $p$, which is the input for the sediment transport model. A transport model is based on the following assumptions:

(a)   the different spatial scales can be handled separately

(b)   the statements for the begin of sediment movement and for the transport rate, known mainly from hydraulic experiments, are transferable to coastal water conditions

(c)   only total load is considered, the transport results from advection

(d)   the driving force is the flow only (wind or tidal induced), not the wave motion

Assumption (a) leads to the concept of separated models for the mesoscale and the macroscale.

According to assumption (b) the Shields-diagram is used, which gives the empirical relation between the critical Froude number for the begin of sediment transport and the grain Reynolds number (Shields, 1936):

$$\left(\frac{\tau_o}{(\frac{\rho_s}{\rho}-1)\,\rho\,g\,d}\right)_{crit} = f\left(\frac{u_* d}{\nu}\right) \qquad (2)$$

at which

$\tau_o = \rho u_*^2$  shear stress
$\rho_s$        density of sediment
$\rho$         density of water
$g$         acceleration due to gravity
$d$         characteristic grain diameter
$u_*$        shear velocity
$\nu$         kinematic viscosity

(2) can also be formulated as a relation between grain diameter and the critical shear velocity.

According to assumption (c) the transport rate is taken from total load formulas, depending on the shear velocity and the critical shear velocity (Graf, 1971; Yalin, 1972). Various formulas where used, which were all similar to the formula of Bagnold (1966):

$$\frac{q_s}{\tau_o \overline{u}} = const + 0.01 \frac{\overline{u}}{w_s} \tag{3}$$

in which

$q_s$     transport rate in quantity per time and length

$\overline{u}$     mean current velocity

$w_s$     settling velocity of the grains

The constant can be taken from an empirical diagram. A deformation of the bottom level h coming from divergences of $q_s$ is calculated by a continuity equation for the sediment:

$$\kappa \frac{\partial h}{\partial t} + \frac{\partial q_s}{\partial x} = 0 \tag{4}$$

Here $\kappa$ is a number expressing the porosity of the sediment (if $q_s$ is given in volume per length and time).
Assumption (d) leads to a restriction of the validity of the model, because wave motion can play an important role for sediment transport near the coast. However, the effect of wave motion causes rather a defacto reduction of the critical shear velocity (respectively an increase of suspended material) than a direct contribution for the horizontal transport. An improvement in this connection will be considered in part 5.

The above formulated transport model contains a number of empirical parameters that have to be determined through comparison with experimental or field data.

## 3. Modelling in the mesoscale

In this scale the vertical current profile is of great importance; therefore the model must contain a discretizing in the vertical direction. For simplification: only one horizontal dimension was considered. Assuming further a constant density of the water, and disregarding the Coriolis force and horizontal exchange, one gets from (1) the following equations ($u = v_1$, $w = v_3$):

$$\frac{\partial u}{\partial t} + u\frac{\partial u}{\partial x} + w\frac{\partial u}{\partial z} - \frac{\partial}{\partial z}\left(A_z\frac{\partial u}{\partial z}\right) + \frac{1}{\rho}\frac{\partial p}{\partial x} = 0$$

$$\frac{\partial w}{\partial t} + u\frac{\partial w}{\partial x} + w\frac{\partial w}{\partial z} \qquad\qquad + \frac{1}{\rho}\frac{\partial p}{\partial z} = 0 \tag{5}$$

$$\frac{\partial u}{\partial x} + \frac{\partial w}{\partial z} \qquad\qquad\qquad = 0$$

The eddy viscosity $A_z$ for the vertical turbulent transport of momentum is calculated by means of a second-order-closure turbulence model (Launder and Spalding, 1972). It has turned out, that in this way the spatial distribution of $A_z$ can be calculated over a rippled bed. The turbulence is characterized by the turbulent kinetic energy k and an other quantity (in the model presented here, this is the dissipation rate of k, which is called $\varepsilon$), which are both calculated from transport equations:

$$\frac{\partial k}{\partial t} + u\frac{\partial k}{\partial x} + w\frac{\partial k}{\partial z} - \frac{\partial}{\partial z}\left(\frac{A_z}{\sigma_k}\frac{\partial k}{\partial z}\right) - \frac{k^2}{\varepsilon}\cdot Prod + c_D\ \varepsilon = 0$$

$$\frac{\partial \varepsilon}{\partial t} + u\frac{\partial \varepsilon}{\partial x} + w\frac{\partial \varepsilon}{\partial z} - \frac{\partial}{\partial z}\left(\frac{A_z}{\sigma_\varepsilon}\frac{\partial \varepsilon}{\partial z}\right) - c_1 k\ Prod + c_2\ \frac{\varepsilon^2}{k} = 0$$

(6)

in which $\quad Prod = 2\left(\frac{\partial u}{\partial x}\right)^2 + 2\left(\frac{\partial w}{\partial z}\right)^2 + \left(\frac{\partial u}{\partial z} + \frac{\partial w}{\partial x}\right)^2$

The empirical constants $c_D$, $c_1$, $c_2$, $\sigma_k$, $\sigma_\varepsilon$ are somewhat universal and are taken unchanged from the literature (Launder, Spalding, 1972). The eddy viscosity results from

$$A_z = \frac{k^2}{\varepsilon}$$

(7)

The efficiency of the system {5, 6} for the calculation of the velocity field was tested by means of the "classical" experimental results of Raudkivi (Raudkivi, 1963). The experiment was carried out in a flume with a rippled bed in a stationary flow ($\overline{u} \approx 30$ cm/s). The computational grid shown in Fig. 2 was used for the calculation.

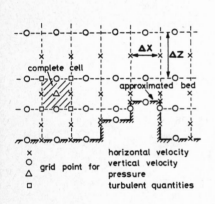

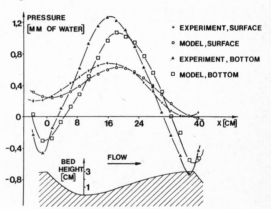

Fig. 2. Computational grid for the mesoscale model

Fig. 3. Comparison of measured and calculated pressure. The measurements are taken from Raudkivi (1963, 1967)

The solution was achieved with an implicit difference procedure. In Fig. 3 we have first the calculated and the measured pressures at the water surface and at the bed.

Fig. 4. shows for various points along the ripple the vertical profiles of the mean horizontal velocity u, the Reynolds stress $-\overline{u'w'}$ and the turbulent kinetic energy k. The agreement between calculated and measured values is quite good. One must keep in mind, that there was no tuning of the empirical constants.

258

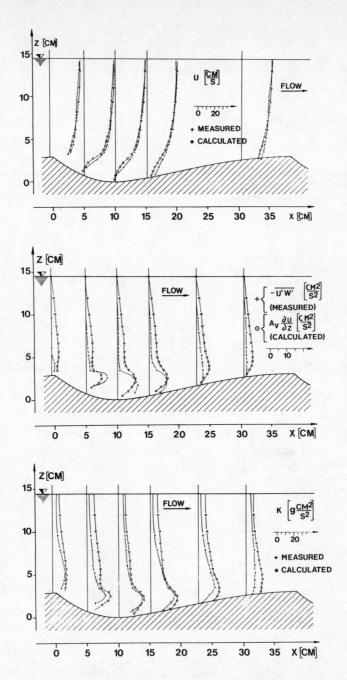

Fig. 4. Comparison of measured and calculated current velocities and turbulent quantities. The measurements are taken from Raudkivi (1963, 1967).
upper picture:  mean horizontal velocity
middle picture: Reynolds stress
lower picture:  turbulent kinetic  energy

The isolines of k (calculated) are given in Fig. 5.

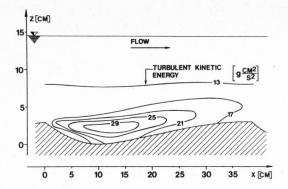

Fig. 5.  Isolines of calculated turbulent kinetic energy

Now a sediment transport model (Puls, 1976) is coupled to the flow
model. First the interactions 1 to 4, shown in Fig. 1, are considered.
The transport model uses the same grid spacing as the flow model and
contains during one time step $\Delta t$ the following three phases of sediment transport (Fig. 6.):

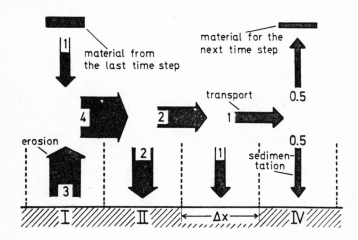

Fig. 6.  Principal structure of transport within one time step. The
         picture shows the fate of 4 sediment units starting from
         interval I.

(a)  Erosion
     a certain amount of sediment is lifted from the bottom:
     $\eta = \text{const.} \cdot (u_*^2 - (u_*)_{cr}^2)$

(b)  Transport
     $\eta$ (and rest material from the last time step) are transported by
     a characteristic "transport velocity" $u_T$; thus the transport rate
     $q_s$ is determined.

(c)  Sedimentation
    a part $q_s/\sigma$ settles down to the bottom ($\sigma$ : path length of the
    material in motion).

The resulting deformation of the bed is the net effect of erosion and
sedimentation.

The quantity $u_T$ is taken from the mean velocities near the bottom. The
shear velocity results from the turbulent kinetic energy near the
bottom (the successful use of calculating $u_*$ from k was shown by Puls
et.al. (1977)):

$$u_* = 0.55 \sqrt{k_{bed}}$$

The so formulated transport model was tested on ripples, that migrated
in the experiment of Raudkivi without changing their shape. Fig. 7.
shows the calculated migration of Raudkivi's ripples. The height remains constant, but there is a little deformation of the upstream
slope. Such small discrepancies must always be taken into account.
What is important, is that the ripple behaviour can be described at
all.

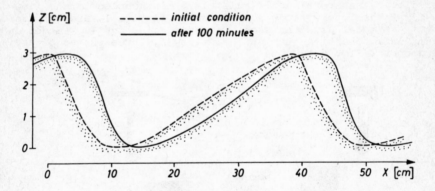

Fig. 7.  Calculated migration of dunes

A further application of the transport model was the deformation of a
sinusoidal bed in a tidal current, which is shown in Fig. 8. The result is the typical shape of dunes in tidal waters: a broad trough and
a narrow crest. This result was encourageing for the calculation of
the development and behaviour of tidal dunes in the field.

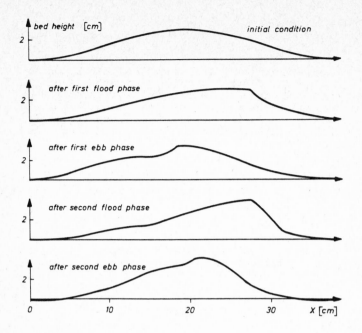

<u>Fig. 8.</u>  Deformation of a sinusoidal bedform by tidal currents
(the flood current goes to the right)


## 4. Modelling in the macroscale

In this scale global transport budgets in the coastal zone and in
estuaries are regarded together with large scaled morphological changes. The detailed shape of the bed is not of interest. Therefore the
vertical component of the flow is not involved; the currents are vertically integrated. The transport processes are simulated in a relatively rough way, which is, however, sufficient for engineering practice.

Vertically integrated velocities $\overline{v_i}$ are introduced into the system (1)
by:

$$\overline{v}_i = \frac{1}{h+\zeta} \int_{-h}^{\zeta} v_i(z) \, dz$$

(h: water depth, $\zeta$: free surface). Further it is assumed that the
water density is constant, the pressure distribution over z is hydrostatic and that the horizontal eddy viscosity is zero. Then one gets
the following system of equations:

$$\frac{\partial \overline{v}_i}{\partial t} + \overline{v}_j \frac{\partial \overline{v}_i}{\partial x_j} + \varepsilon_{ij} \overline{v}_j = -g\frac{\partial \zeta}{\partial x_i} - \frac{\tau_i^b}{h+\zeta} + \frac{\tau_i^s}{h+\zeta}$$

$$\frac{\partial \zeta}{\partial t} + \frac{\partial}{\partial x_j} ((h+\zeta)\overline{v}_j) = 0 \qquad (i,j = 1,2)$$

(8)

Here the last two terms of the equation of motion charaterize the tangential stresses by bottom friction and wind. These stresses are the product of boundary conditions for the vertical integration.

The numerical solution of (8) with the help of a horizontal grid gives the spatial and temporal distribution of the water level $\zeta$ and the velocities $\overline{v}_i$. The coupling of the convective transport model to the flow model is the same as with the mesoscale model, i.e., by using the common relations (2) and (3). The shear stress or the shear velocity in these relations are transformed to vertically integrated mean velocities

by $u_{*_i} = 0.07\ \overline{v}_i$ \hfill (9)

Bagnold's transport formula is written in computer compatible form:

$$q_{si} = \left\{ \begin{array}{ll} q_{si}^o\ (\overline{v}_j\ \overline{v}_j - \overline{v}^2_{crit})^\alpha\ \overline{v}_i & \text{for } |\overline{v}_j| > \overline{v}_{crit} \\ 0 & \text{else} \end{array} \right. \tag{10}$$

$q_{si}^o$ and $\alpha$ are constants, that are estimated from field data by calibration. The critical velocities $\overline{v}_{crit}$ are from Shield's relation (2) and from (9).

The so formulated two dimensional macroscale model was successfully applicated to the global sand transport in the North Sea (Sündermann and Krohn, 1977).

In the following the results of a one dimensional application of the macroscale model to the sediment transport in the German tidal river Eider will be presented. This river had been dammed up in 1936 at about 30 km upstream of the mouth in order to shorten the dyke line. As a consequence there was a strong silting up of the river (Fig. 9.), which leaded to a drastic narrowing of the sectional area within 10 years. As a consequence there was an obstruction of navigation and of the flood wave.

The horizontal velocity $v_1$ and the water level were calculated with the one dimensional version of system (8). Fig. 10. shows for one point (near km 10) the time sequence of these values before and after the damming up. For demonstrating the correctness of the calculations, the measured curves were added. There was no calibration for the situation after the construction of the dam, because it was only the capability to predict, which was to be tested. One can see, that before the damming up, the upstream and the downstream transports were obviously balancing each other, because the current velocities for the flood and the ebb were similar. After the damming up, the flood current is much stronger than the ebb current, which causes the silting up.

Fig. 11. shows a provisional result of the transport model with equation (10). Near the mouth of the river the silting up shown in Fig. 9. is reproduced qualitatively. The investigation about this subject is continued.

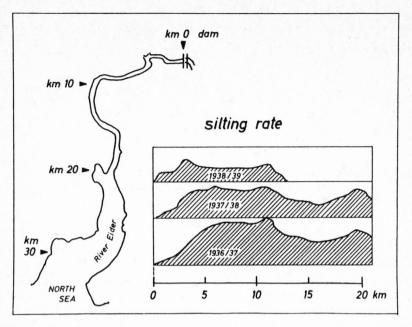

**Fig. 9.** Historical silting up of the river Eider. Sedimentation rate along the bed for the first three years after damming up. (measured in volume per unit length of the river axis)

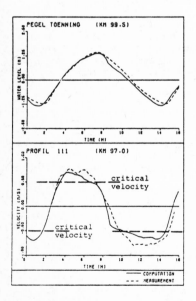 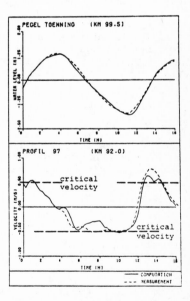

**Fig. 10.** Computed and observed water elevation and stream velocity at km 10 before (left) and after (right) the damming up. For demonstration an arbitrary critical velocity has been assumed.

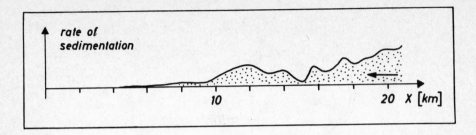

Fig. 11. Computed silting up of the river Eider after half a year
after damming up (measured in volume per unit length of
the river axis)

## 5. Conclusion

The present investigations are to be considered as first steps towards a quantitative treatment of sediment transport by finite numerical techniques. The demonstrated results showing the principal
agreement of computed and observed tendencies are encouraging further
investigations with improved models. Main topics in this connection
would be a parametrisized interaction with surface waves following
the concept of Madsen and Grant (1976), the consideration of natural
sediment fractions and the application of new numerical techniques.
The developed models must be permanently tested against observations.

From an engineering viewpoint, the main aim is not to achieve a theoretically accomplished model simulating in detail the governing physics
(if ever that is possible), but to establish a practical algorithm,
which gives rough quantitative estimates of the sediment budget without great effort, and which can be used also for prognostic purposes.

## Acknowledgement

This research was sponsored by the Deutsche Forschungsgemeinschaft
(German Science Foundation) through the Sonderforschungsbereich 79
(Coastal Engineering). We thank Mr. R. Klöcker for programming the
models of the river Eider.

References

Bagnold, R.A.: An approach to the sediment transport problem from
general physics. U.S. Geol. Survey, Prof. Paper 422-J (1966)
Graf, W.H.: Hydraulics of sediment transport, McGraw-Hill (1971)
Launder, B.E., Spalding, D.B.: Mathematical models of turbulence,
Academic Press, London (1972)
Madsen, O.S., Grant, W.D.: Sediment transport in the coastal environ-
ment. Ralph M. Parsons Lab. for water resources and hydrodynamics,
MIT, Cambridge/Mass., Rep. No. 209, p. 105 (1976)
Puls, W.: Ein mathematisches Modell für die Wechselwirkung Strömung-
Feststofftransport-Sohldeformation in einem Gerinne, Dipl.Arbeit,
Universität Hamburg (1976)
Puls, W., Sündermann, J., Vollmers, H.: A numerical approach to solid
matter transport computation. Proc. 17th congress of the IAHR,
Baden-Baden, Vol 1, 129-135 (1977)
Raudkivi, A.J.: Study of sediment ripple formation. ASCE, HY6 (1963)
Raudkivi, A.J.: Loose boundary hydraulics. Pergamon Press, Oxford
1st edition (1967)
Shields, A.: Anwendung der Ähnlichkeitsmechanik und Turbulenzforschung
auf die Geschiebebewegung. Mitteil. Preuss. Versuchsanstalt
Wasser-, Erd-, Schiffsbau, Berlin, no. 26 (1936)
Sündermann, J., Krohn, J.: Numerical simulation of tidal caused sand
transport in coastal waters, Proc. 17th congress of the IAHR,
Baden-Baden, Vol. 1, 173-181 (1977)
Yalin, S.: Mechanics of sediment transport, Pergamon Press (1972)

B. Le Méhauté

# An Introduction to Hydrodynamics and Water Waves

Springer Study Edition

1976. 231 figures, 12 tables. VIII, 323 pages
ISBN 3-540-07232-2

*Contents:* Establishing the Basic Equations that Govern Flow Motion. – Some Mathematical Treatments of the Basic Equations. – Water Wave Theories. – Appendices: Wave Motion as a Random Process. Similitude and Scale Model Technology.

"...The declared aim of the book is to provide a text based on a set of lecture notes for engineering students, that introduces the mathematical aspects of fluid mechanics with explanations of physical meaning to help practising engineers to read mathematical texts and keep pace with new developments reported in scientific journals... The text is divided into three main parts concerned, respectively, with the derivation of the basic equations of fluid motion, mathematical treatment of the equations, and water wave theories... Examples of clarity include well illustrated accounts of the distinctions between streamlines, stream tubes, streak lines and particle paths and of the significance of dilatation, shear and rotation in interpretating the kinematic equations. I specially commend the many well organised tabular summaries of key details and formulae which students will surely find very helpful and which offer convenient working references for practising engineers..."

*N. Hogben in: Nature*

Springer-Verlag
Berlin
Heidelberg
New York

J. Pedlosky

# Geophysical Fluid Dynamics

1979. 180 figures, 2 tables. XII, 624 pages
ISBN 3-540-90368-2

**Contents:** Preliminaries. – Fundamentals. – Inviscid Shallow-Water Theory. – Friction and Viscous Flow. – Homogeneous Models of the Wind-Driven Oceanic Circulation. – Quasigeostrophic Motion of a Stratified Fluid on a Sphere. – Instability Theory. – Ageostrophic Motion. – Selected Bibliography. – Index.

This comprehensive and lucidly written introduction to the theory of ocean and atmospheric movement dynamics blends physical and intuitive reasoning with mathematical analysis. The book combines approximations of the basic fluid dynamical equations required for progress with the precise analysis necessary to make the progress meaningful. Emphasis is placed on the systematic derivation and application of quasigeostrophic theory to central problems of oceanography and meteorology. Designed for students and scientists, this book lays the basis for understanding and conducting modern research in oceanography and meteorology.

# Oceanic Fronts in Coastal Processes

Proceedings of a Workshop Held at the Marine Sciences Research Center, May 25–27, 1977
Editors: M. J. Bowman, W. E. Esaias

1978. 84 figures. IX, 114 pages
ISBN 3-540-08823-7

**Contents:** Introduction and Proceedings: Introduction and Historical Perspective. Proceedings of the Workshop. – Contributions by Participants: Frontal Dynamics and Frontogenesis. Advection-Diffusion in the Presence of Surface Convergence. Shallow Sea Fronts Produced by Tidal Stirring. Prograde and Retrograde Fronts. Physical Aspects of the Nova Scotian Shelf-Break Fronts. Biological Aspects of the Nova Scotian Shelf-Break Fronts. Headland Fronts. Estuarine and Plume Fronts. Crossfrontal Mixing and Cabbeling. – Appendices.

"An ensemble of articles that represent three days of cloistered, intense discussion by 12 acknowledged experts in the field of coastal oceanic fronts. The effort is an excellent summary of the state of the art in 1977. The book presents the results of the workshop, together with nine contributions by the participants. There is no attempt to present a comprehensive review of the relevant literature, instead the contributions give a multifaceted view of the research in coastal fronts. …Recommended for professionals and graduate students specializing in physical oceanography and in meteorology."

*Choice*

# Springer-Verlag Berlin Heidelberg New York